Beyond Disruption

Technology's Challenge to Governance

George P. Shultz, Jim Hoagland, and James Timbie
Editors

HOOVER INSTITUTION PRESS

Stanford University | *Stanford, California*

With its eminent scholars and world-renowned library and archives, the Hoover Institution seeks to improve the human condition by advancing ideas that promote economic opportunity and prosperity, while securing and safeguarding peace for America and all mankind. The views expressed in its publications are entirely those of the authors and do not necessarily reflect the views of the staff, officers, or Board of Overseers of the Hoover Institution.

www.hoover.org

Hoover Institution Press Publication No. 688

Hoover Institution at Leland Stanford Junior University,
Stanford, California 94305-6003

First printing 2018
24   23   22   21   20   19   18   7   6   5   4   3   2   1

Manufactured in the United States of America

The paper used in this publication meets the minimum Requirements of the American National Standard for Information Sciences—Permanence of Paper for Printed Library Materials, ANSI/NISO Z39.48-1992. ∞

Cataloging-in-Publication Data is available from the Library of Congress.
ISBN: 978-0-8179-2145-3 (pbk. : alk. paper)
ISBN: 978-0-8179-2146-0 (epub)
ISBN: 978-0-8179-2147-7 (mobi)
ISBN: 978-0-8179-2148-4 (PDF)

CONTENTS

PREFACE

The contents of this book were discussed at a special conference on governance and technological change at Stanford University's Hoover Institution on September 27–28, 2017. The conference was dedicated to our late Hoover Institution colleague Sid Drell, an eminent theoretical physicist who was universally admired.

Sid played an indirect but important role in the effort by President Ronald Reagan to bring about reductions in our nuclear arsenals. Key advisers—Paul Nitze and Jim Timbie (my co-editor for this book)—often talked about ideas attributed to Sid Drell. Sid's Hoover office was a beehive of talk—yes, about nuclear weapons issues and national security but also about a wide range of other subjects. Visitors such as his physics friends, Raymond Jeanloz (UC–Berkeley) and Chris Stubbs (Harvard), pitched in. Their chapter appears in this book.

Sid was an outstanding physicist with a great mind. He also had a wonderful capacity for friendship. Often, Jim Mattis, who was a Hoover Fellow for three years or so before becoming US secretary of defense, would join our conversations. Jim consulted with Sid on material he was writing, and Jim soon came to admire and respect Sid as much as we all did. Jim was unable to attend the conference in Sid's honor, but he wrote the following letter to be read during the proceedings:

There could be no finer inspiration for this conference than the man we called Sid.

Dr. Sidney Drell was a giant in voice, intellect, spirit and . . . laughter. His example of unbending ethics, rigorous intellect, and irrepressible joy was matched by his love of family, physics, music . . . and life itself. Although our formative experiences came from opposite sides of the earth, I could have searched the world yet nowhere could I have found a finer friend.

Sid Drell was concerned about the issues discussed at our conference, and he encouraged the idea of holding such a meeting at the Hoover Institution. So, in a very real sense, our conference was a reflection of Sid's wisdom. That wisdom will continue to be present as we ask ourselves—when confronted with problems of various kinds—what kind of advice Sid would have given us.

—George P. Shultz

INTRODUCTION

Jim Hoagland

Since early humans created the first stone tools, technology has brought great benefits accompanied by significant disruption and peril for its inventors and society. The first decades of the twenty-first century have brought a surfeit of such advantage and challenge.

But this time, technological change is both ubiquitous—arriving nearly simultaneously in all parts of the globe—and immediate, shrinking time as well as space. Decision-making, the hallmark of governance, is being severely disrupted. So are the political systems and industrial workplaces of the world's democracies. The goal of this book, and of the conference at the Hoover Institution that gave rise to it, is to examine a still unfolding era of technological change that is unprecedented in scope, pace, and effect.

This era's dangers echo those of the atomic age, which persist in newly menacing forms. Nations must now wonder if their nuclear weapons—and power plants—are suddenly vulnerable to cyberattack. Other challenges swirl through the digital information revolution, the rapid spread of artificial intelligence (AI) and additive manufacturing, and the biosciences that heavily influence the nature of human life itself. This storm of change appears to be poorly understood by the world's citizens and insufficiently anticipated by their governments. The chapters and discussion excerpts that follow attempt to deepen that popular understanding and to expand the burgeoning debate about remedies that local and national governments can pursue in seeking to foster the best—or at least avoid the worst—probable outcomes.

Technology—defined by the Oxford English Dictionary with stunning simplicity as "the application of scientific knowledge for practical purposes, especially in industry"—no longer is primarily about replacing older machines with newer machines or manual labor with robotic assembly lines. Technology today is moving deeply into performing and even managing cognitive tasks that were once the exclusive domain of humans. This threatens not only to disrupt the livelihoods of managers, and others who work with the mind and senses but also to corrupt the form and content of language and communication.

People everywhere can more easily learn what is going on, and they can easily communicate and organize to support or oppose what others propose. In this new age of national and international transparency, political and intellectual diversity is an established fact of life. It cannot be easily ignored or suppressed. Governments must learn to govern over diversity in all its forms while advancing its positive contributions to civilization.

But the new electronic forms of social networking foster polarization, constraining governments' ability to function. An unintended consequence of the communications revolution has been to enable the targeting of large numbers of people to receive false and misleading information that reinforces their beliefs and biases. This helps fragment debate into man-made canyons of hate speech, bullying, and denigration. Public opinion surveys suggest that this aspect of the communication revolution contributes significantly to a growing popular mistrust of national institutions, political leaders, and fellow citizens.

The chapters that follow explore how best to maintain the significant benefits that the new technologies provide while mitigating their dangers. One clear common theme that emerges is that the new disruptions pose similar perils for virtually all societies—and must be dealt with on a global basis. The responses range from developing ways to protect against or deter cyberwarfare to resisting the galloping spread of infectious diseases. The global commons of cooperative international institutions created out of the ruins of World War II is in urgent need of rehabilitation and redirection to keep pace with twenty-first-century technological change.

Time is in many ways the critical dimension of the fourth industrial revolution that several of our authors say is now under way. During the first three transformations—sparked successively by the steam engine, electricity, and computer science—governments were able to adapt gradually and mitigate disruptions. The move of agricultural workers into city-based employment took place over decades and could be cushioned by gradual change in social policies and educational structures.

Today, the swiftness with which software programmers can use powerful algorithms to eliminate or create jobs or to penetrate bank records has significantly compressed the ability of governments and corporations to keep up and adapt. This adds to a sense of modern society's fragility, with both the affluent and struggling feeling targeted.

This onslaught of challenges is stirring a broad response that the authors seek to identify and encourage. As artificial intelligence occupies a growing place in industry, workers, businesses, and governments are creatively seeking ways to make the reeducation and reskilling of workforces a constant, lifelong endeavor rather than once-in-a-lifetime episodes experienced on a campus or in a job-training course.

In "Technological Change and the Workplace," James Timbie explores the tensions being created by this new wave of job destruction and creation. It will ultimately "increase national wealth and income" through increased productivity, he notes. But its effects will be felt unevenly, with "low- and middle-income workers without a college education" feeling "the most pressure on wages and employment."

That there is a mismatch between the skills possessed by workers in these vulnerable positions and the skills needed in a new AI-centered economy is suggested by the fact that US employers report they cannot currently fill six million open jobs. And demographic trends suggest that slower growth in the working-age population will inhibit productivity gains in years ahead. These problems must be urgently addressed by individual and community action including support for community college training programs and coaches to help displaced workers transition to new jobs.

Timbie also examines the effect of "big data," information collected from internet activities and—increasingly—through unperceived sensors contained in the "internet of things." A better outcome is likely if the tech industry provides consumers with more effective, easily understandable standards of privacy. Failure to do so could provoke government action.

T. X. Hammes's chapter 2 looks over the horizon of US national borders and of contemporaneous events to describe an emerging landscape of "deglobalization." That is, the transportation networks, offshore labor forces, and supply chain systems that fed the significant expansion of world trade and investment flows in the past half-century are being disrupted by additive manufacturing techniques built around 3D printing; the surge of automation, which increasingly combines the work of humans and relatively low-cost robots; and new sources of cheaper energy. The United States is in fact leading the way at the outset of this rebirth of local manufacturing.

The growing availability of relatively cheap, powerful, and autonomous weapons systems such as unpiloted drones will also change the nature of warfare, Hammes submits. As smaller countries, terrorist groups, and individuals obtain even more access to greater destructive capabilities, the Pentagon must move away from strategies built around expensive, sophisticated weaponry.

Hammes also notes that the projected decline in world trade will add to existing political pressures for Americans to look inward.

The cumulative effect of such changes in military technology is to undermine the existing concepts of strategic stability among great powers based on "diplomacy, deterrence, and direct military action," write Raymond Jeanloz and Christopher Stubbs. National leaders must now determine in a matter of minutes whether and how to respond to reports of an imminent or actual launch of enemy missiles. And their ability to determine with accuracy who may have launched a cyberattack on national infrastructure is severely limited. The authors encourage building greater resiliency and accuracy into command-and-control systems and warn against incorporating into them still poorly understood features of AI. Instead, they urge that the resiliency of national infrastructure be hardened to reduce such fragility.

Their warnings are reinforced by James O. Ellis, Jr., who says the United States today confronts "a threat landscape unlike any we have faced." He then details the vulnerabilities of a "global operating system" that dominates our hyperconnected world. In a resource-constrained, threat-rich environment, national authorities must prioritize the risks the nation faces. It is necessary "to measure risk, minimize the risk to the extent possible, manage the risk that inevitably remains and . . . be prepared with a mitigation plan when the next crisis materializes." A key component of this lies in having cyberprotection that is "designed in, not bolted on" to new operating systems.

The health of the planet and the humans who inhabit it are other areas of dangerous disequilibrium created by human actions, Lucy Shapiro and Harley McAdams observe. "Rapid global movements of formerly local pathogens and their vectors" are creating a redistribution of infectious diseases to new regions, they write. With global warming, tropical diseases are moving north. Diseases spread by mosquitoes are of particular concern. "These changes affect the health of people, ocean life, and the animals and plants that are our food sources. . . . Since the 1980s, the annual number of epidemics across the globe has tripled. . . . The death of reefs, from the Australian Great Barrier Reef to the reefs in the Caribbean, is causing a catastrophic disruption in the global food chain."

Balanced against this grim prospect are life-improving medical advances in diagnostic and treatment tools. And the promise of genetic engineering and editing "expands the tool kit for basic research in living systems."

A determining factor in the race between good and bad outcomes of technological change will be the quality of scientific education and research fostered by governments at national and local levels. Shapiro and Adams report cause for concern for Americans:

Increasingly, "the trained and talented research scientists who can develop the solutions to these problems . . . will have to come from the international community. . . . US fifteen-year-olds ranked twenty-first in science and twenty-sixth in math" on international standardized tests administered in 2012 among the thirty-four member nations of the Organisation for Economic Co-operation and Development. Beyond that, Shapiro and McAdams report, over the past twenty years, the United States has "moved from first to tenth place in R&D investment as percentage of GDP among industrialized nations." Providing citizens with education that enables them to live and work productively is one of the most essential tasks of governance. Reversing the current downward trend is therefore a major step in enabling the nation to deal with the disruption and disrepair that otherwise will threaten.

Looking back into history to glimpse the future also yields informative perspectives on today's problems. Governments must face up to the profound transformation of the global power structure "by the proliferation of distributed networks" across national frontiers. Those horizontal networks now challenge vertical, hierarchical organizations, including the nation-state and the varying concepts of "world order," Niall Ferguson writes. The information technology revolution "was almost entirely a US-based achievement" that gave the rest of the world two options: "capitulate and regulate or exclude and compete." Europe chose the first, China the second, he notes.

In his conference presentation, Ferguson offered the 2016 US presidential election as a prime example of how horizontal networks are changing the world. The Trump campaign harnessed the networks of Silicon Valley, "to the dismay of the people who owned and thought they also controlled the networks," while Russia's intelligence network "launched a sustained assault on the American political system." His critical point is that Moscow was undeterred from using what Ferguson calls "Cyberia" tactics.

William Drozdiak exposes the doctrine behind this Russian strategic cybercampaign to disrupt US and European democracies by disseminating "fake news" and other propaganda through social media networks. But he concludes that—in France, Germany, and Estonia, at least—stiff public and private resistance to Kremlin interference caused the Russians to back off. Coupling demonstrations of political will and unity with expanding media literacy courses for school-age populations may be among the solutions to problems created by the internet.

America's system of checks and balances—built by the Founders around concepts of a representative republic rather than a direct democracy—has been profoundly disrupted by the continuing communications revolution, as David M. Kennedy demonstrates in detail in the penultimate chapter of this volume. Presidents can now offer their campaign or programmatic slogans such as "a New Deal" or "Make America Great Again" directly to electorates, bypassing political parties as well as factual authorities such as the press. A "plebiscitary" presidency has brought a century-long "process of disintermediation to an extreme conclusion" and set it "free from editorial curating or fact-checking or even the protocols of civil speech."

This concern is shared by Charles Hill in the concluding chapter: "The emergence of a certain type of modern state, at once ideological and dictatorial . . . has given propaganda a wider scope and intensity. Autocratic regimes see these communicative breakthroughs as new ways to increase their powers over their peoples."

Moreover, he notes that technology's intense compression of time is a creeping danger for civilization, which depends on society having the time and the ability to contain harmful eruptions of emotion or malice. And he explores how "the volcanic eruption of e-communication in the last few years has brought an exceptional array of challenges to governance in the Arab-Islamic realm," where the spoken word has traditionally dominated written expression.

The authors do not pretend that they have found easy or quick fixes for the daunting challenges that technology and diversity present to governance in the twenty-first century. But their work as a group underlines the urgent need for creative thinking and sustained focus by world leaders and their populations on new forms of international cooperation. These would identify and proscribe the most dangerous and destructive practices of cyberwarfare, the growing threats to space-based systems, and surveillance and harassment of people by electronic means, much in the way moral strictures against chemical and biological warfare have been translated into international pacts. Existing international organizations may need to be reorganized, revitalized, or overhauled to achieve this, some authors suggest, to provide new concepts and instruments for global stability and progress.

Change has become a constant in a world that has in a relatively few years seen huge numbers of people lifted out of grim poverty, superpower nuclear arsenals sharply cut, and education spread to the most remote corners of the earth. Our purpose is to focus and keep attention on what needs to be done to make sure this positive progress continues and flourishes.

1

TECHNOLOGICAL CHANGE AND THE WORKPLACE

James Timbie

Part I. Advancing Technology and Its Impact

The computation and communication technologies that enabled the information revolution continue to advance rapidly, promising benefits and efficiencies but also raising questions concerning displacement of workers, inequality, and privacy. This chapter outlines further advances in automation and artificial intelligence (AI) and in the scale of collection and exploitation of data that can reasonably be foreseen over the next ten to fifteen years. These technologies are in widespread use today. But impressive new applications of artificial intelligence (such as self-driving vehicles) and dramatic increases in the amount of data collected by the "internet of things" (IoT) and stored in the cloud are on the near horizon. There is potential for further advances in health, safety, and productivity—and also for further disruption of working and personal lives.

Artificial Intelligence[1]

Recent advances in software and hardware, combined with the availability of large sets of digital data, have enabled the development of machines with the ability to sense their environment, learn by trial and error, solve problems, and take action. It is no longer correct to say that machines only do what they are programmed to do. Machines can now be trained to learn from examination of large amounts of data. Machine learning is the aspect of artificial intelligence responsible for much of the disruption of the workplace.[2]

The twenty-first century has seen unanticipated progress in machine learning that has enabled dramatic advances in practical applications such as speech recognition and language translation. In the near future, we will see autonomous vehicles, better diagnostics for the sick, and better prevention strategies for the healthy. The combination of artificial intelligence and advancements in other technologies such as robotics and 3D printing ensures that the rate of change in many industries will continue to be swift, with accompanying social and economic consequences. This chapter begins with a summary of what can now be expected in certain sectors.

Transportation[3]

Advances in sensors and machine learning have accelerated the development of self-driving vehicles, which are likely to become widely available in the next few years. Autonomous long-haul trucks are expected to be introduced soon; autonomous cars and trucks for city use will come later, as the technology is developed for safely navigating the more complex and less predictable urban environment. Self-driving cars promise to be substantially safer and could transform commuting into an opportunity for constructive activity. Self-driving trucks and autonomous robots will reduce transport and delivery costs. Autonomous vehicles and related transportation services could reduce incentives to own cars, especially in cities, and encourage new forms of public transportation based on smaller vehicles that transport people on demand from point to point. These safety, convenience, and efficiency benefits will come at a cost to those currently driving trucks, buses, taxis, and ride-sharing services.

Health Care[4]

Artificial intelligence technologies have the potential to improve health outcomes and quality of life through better clinical decisions, better monitoring and coaching of patients, and prevention of disease through early identification of possible health risks. Machines could learn which practices and procedures lead to the best outcomes by analyzing vast amounts of data collected in electronic medical records of millions of patients. They could also identify unintended negative effects of procedures and drugs. Machines trained by correlating electronic medical images with data on patient outcomes will enhance the accuracy of interpretation of medical images. The combination of human physicians and machine intelligence will enhance the accuracy of diagnoses of problems and recommendations for therapy and further tests. Additional sources of personal health information from personal fitness devices and social media, for example, and information on individual genomes will support more personalized diagnosis and treatment, along with an emphasis on prevention rather than cure.

We will continue to want a human physician to evaluate the output of machine intelligence in making clinical decisions and recommendations for treatment. The doctor would convey the outcome to the patient and help the patient understand and accept it. The role of nurses, who interact with patients, communicate with them, and make them comfortable, is less susceptible to disruption by automation in the near term.

Education[5]

Machines will not replace teachers. But over the next ten to fifteen years, the use of systems based on AI technologies in the classroom and in the home will expand substantially. Interactive machine tutors are being developed to help educate students and train workers in a variety of subjects, providing personalized coaching and support and monitoring progress. Online courses have promise for providing personalized interaction with students at all levels on a large scale, exposing students to courses that have proved successful and allowing them to work at their own pace using educational techniques that work for them. Current experimentation and online courses are producing feedback data that will allow the developers of educational systems to learn what works and to improve, including finding the best mix of machines and teaching assistants to provide support. An important application of online courses and intelligent tutoring systems is likely to be the retraining of workers and lifelong learning. Large-scale, personalized adult education and training can be part of the solution to the disruption of the workplace by changes brought about by advancing technologies.

Manufacturing[6]

For decades, manufacturing in many industries has moved offshore, driven by low labor costs, efficient freight systems, and trade agreements. Disruption of industries and loss of skilled, well-paying jobs have contributed substantially to the problems of governance. But globalization may have peaked; trade as a fraction of GDP is now declining. The combination of artificial intelligence, robotics, and 3D printing promises further fundamental changes in the way things are made, leading to production of goods, services, energy, and food close to the consumer. The falling costs and increasing capabilities of 3D printing, with its inherent ability to customize each item at no additional cost, will allow production of consumer goods and industrial products built to order for each individual customer. Hospital supplies and parts for cars, trucks, and aircraft can be produced when and where they are needed, rather than stockpiled. With AI, advanced robotics, and 3D printing technologies reducing labor costs and increasing quality and customization, the advantages of manufacturing in countries with low labor costs will be reduced, while the advantages of production of made-to-order products near the customer will grow. As one line of argument goes, "With the cost of labor no longer a significant advantage, it makes little sense to manufacture components in Southeast Asia, assemble them in China, and then ship them to the rest of the world when the same item can either be manufactured by robots or printed where it will be used."[7] On shoring is the likely trend for the next ten to fifteen years, but the associated new jobs will be different from those that were lost to offshoring.

The key factor that is coming out of this is cost of labor advantages are disappearing. Industrial robots are cheaper than Chinese labor.   —T. X. Hammes

Employment and the Workplace[8]

The list of industries where automation and artificial intelligence will change the workplace in fundamental ways is long and diverse, including:

- Health care (automated diagnostics, image interpretation, robotic surgery, patient monitoring, risk assessment, and disease prevention)
- Transportation (autonomous cars, trucks, and taxis; monitoring of aircraft engines)
- Law (pretrial discovery)
- Call centers (voice recognition and responses)
- Education (interactive tutors, online courses)
- Software (machines that write and debug software)
- Logistics (automated warehouses, sensors for supply chain management)
- Agriculture (autonomous vehicles, crop and animal monitoring, local indoor farms)
- Elder care (automated transportation, monitoring, personalized health management, service robots)
- Manufacturing (automated production lines of all kinds)

In all of these endeavors, large numbers of workers now onshore and offshore will be displaced by more efficient machines. In contrast to the nineteenth-century mechanization of manual tasks and the twentieth-century offshoring of routine tasks, twenty-first-century machines will be moving into a wide range of cognitive tasks that until now have been reserved for humans, including professional services. One result will be less expensive, better quality, and more customized goods and services, plus an improved standard of living. Another result will be loss of employment for workers in a broad range of skill and income levels.

A study by Frey and Osborne suggests that 47 percent of workers are in occupations with a high probability of displacement by automation.[9] They conclude:

> While computerization has been historically confined to routine tasks involving explicit rule-based activities, algorithms for big data are now rapidly entering domains reliant upon pattern recognition and can readily substitute for labor in a wide range of non-routine cognitive tasks. In addition, advanced robots are gaining enhanced senses and dexterity, allowing them to perform a broader scope of manual tasks. This is likely to change the nature of work across industries and occupations.[10]

They find workers in service industries to be highly susceptible to automation, as well as workers in transportation and logistics, office and administrative support, and production. Machine learning is even assuming some of the tasks of software engineers.

As in the past, new industries and new jobs will be created. The number and nature of these new jobs are difficult to foresee. Certain tasks will become more important, creating opportunities for expansion, and new categories of employment could be created. The net effect on the total number of jobs is difficult to predict. What is clear is that a substantial fraction of the workforce may lose well-paying "cognitive" jobs to automation, perhaps more over time than the well-paying factory jobs lost to globalization.

The recent evolution of chess may provide a hint about the future. After a long period of development of hardware and software, a computer defeated the best human chess player, Gary Kasparov, in a well-known match in 1997. Today, a laptop computer with off-the-shelf software can play as well. Now there is a new game called free-style chess, in which a human player can draw upon machine support.[11] For successful players, the human provides the strategy and uses a variety of machines to explore tactics and consequences of potential moves. The human-plus-machine combination is widely considered to play at a higher level than either humans or machines, and a human-plus-machine combination can defeat any human or any machine.

This lesson may be broadly applicable, suggesting that the best results will come from humans supported by intelligent machines—a combination of a doctor and a machine, a teacher and a machine, and so on. In the workplace of the near future, humans will do jobs (or portions of jobs) that machines do not do well, and work with machines in areas where machines have advantages. Davenport and Kirby describe future workplaces where machines and humans work together, the machines doing the computational work they do best, augmenting the humans who see the big picture and have interpersonal skills.[12] One example would be insurance underwriting, where machines make detailed risk assessments and premium calculations for each application for insurance, and humans address exceptional cases, manage the company's overall risk profile, and communicate with individuals whose applications were denied.

It is not clear whether the number of jobs created and jobs retained in modified form will match the number of jobs lost to artificial intelligence, automation, and robotics. Those who have studied this question are closely divided.[13] Historically, over the two hundred years since the Luddite rebellion, gains in productivity have, over time, led to new jobs in new industries. That could continue. Or this time could be different.

What is clear is that in the near term, large-scale disruption of the workplace will continue and probably accelerate. In contrast to the twentieth century, where job loss was concentrated on middle-income, mid-level skill occupations, the current advances in artificial intelligence and automation in the twenty-first century will affect workers at all skill and income levels (high and low as well as middle), including some well-paying cognitive jobs.

I heard someone say the other day that in the factory of the future we're going to have two living creatures: one a man and the other one a dog. The dog's job is to keep the man from touching anything, and the man's job is to feed the dog.   —Sam Nunn

---

Inequality[14]

For the country as a whole, the advance of technologies that exploit artificial intelligence and automation will likely continue to increase national wealth and income, but these benefits will be distributed unevenly. Some workers performing routine tasks at all skill and income levels will be adversely affected. But low- and middle-income workers without a college education will feel the most pressure on wages and employment. This is a continuation of a well-established trend. Median household income has not risen significantly since 1999, even as GDP has grown 38 percent. All of the gains in income have gone to the upper end (table 1.1).

TABLE 1.1   US Household Income Shares, 2016

	1991		2016
Bottom 20% of households	3.8%	↘	3.1% of total household income
Top 20% of households	46.5%	↗	51.5%
Top 5% of households	18.1%	↗	22.5%

Source: Jessica L. Semega, Kayla R. Fontenot, and Melissa A. Kollar, "Income and Poverty in the United States: 2016," US Census Bureau, Current Population Reports, P60-259 (Washington, DC: Government Printing Office, September 2017).

Other metrics—including median income per capita, total wealth, and life expectancy—also demonstrate growing inequality.[15] The spread of automation contributes to this growing inequality in wealth and income. It is easy to see how this happens. Brynjolfsson and McAfee give the example of TurboTax, a provider of tax preparation software.[16] Many customers find it cheaper, quicker, and more accurate than having tax preparers produce their returns. TurboTax has therefore created a great deal of value for its users. The small cadre who created TurboTax has benefited handsomely. But a much larger number who earned their living as tax preparers now find their jobs and income threatened. Replicating this example throughout the economy, new technology adds to the nation's GDP and standard of living, concentrates new wealth in a small number of entrepreneurs and their skilled employees, and threatens the livelihoods of a larger number of displaced workers.

The failure of median incomes to rise as GDP grows contributes to the widespread perception among low- and middle-income citizens that the present economy does not work for them.

Is it us or is it the world? How much of this is because of the way we're doing things, and how much of this is unavoidable because of technological change?   —Burton Richter

Big Data

The recent and surprising advances in artificial intelligence for practical applications are based on the application of innovative processing power and software to very large sets of digital data. For example, machines have recently become quite proficient at translation by comparing vast amounts of digital text collected from many sources. These translation systems are not programmed on grammar, syntax, or spelling; they are trained by examining very large amounts of text in various languages, enabling them to learn how a phrase in one language correlates with the corresponding phrase in another. Similarly, machines with no knowledge of biology are trained to interpret medical images by finding correlations between digital images and medical outcomes, based on lots of data from very large numbers of cases.

"Big data" is quantitatively very big. Google processes over 3.5 billion search queries each day and saves information on each one. Facebook uploads more than 300 million photographs each day. This enormous growth in scale results in a qualitative change as well. Collection of so much information—all or nearly all of the information on a subject—has facilitated a transformation away from drawing statistical inferences from a sample of data to drawing deeper, more detailed, and more reliable conclusions by examining all the data, not just a sample. Given a very large data set, machines can learn to find patterns and correlations and make reliable predictions without considering the physical or biological processes involved. As the data set continues to grow, the predictions get better. The role of vast amounts of data in the success of artificial intelligence is so central that the entire field is coming to be called "data science," two key components of which are data collection and machine learning to draw conclusions from the data and make predictions.

An illustrative example of the constructive combination of machine learning and big data is Google's use of search queries to detect and track seasonal flu outbreaks.[17] Google researchers programmed machines to examine records of hundreds of billions of search queries between 2003 and 2008, along with historical information collected by the US Centers for Disease Control and Prevention (CDC) on patient visits to doctors reporting flu-like symptoms. Google's automated system was given no information about influenza or how it is treated, but was able to identify forty-five search queries that are correlated with a flu outbreak in a region. The machines looked for and found correlations in a vast quantity of data, succeeding where earlier attempts using less data had failed. Google's system provided accurate estimates of the spread of flu in near real time, without the one- to two-week lag of traditional CDC reporting. This technique was soon put to practical use when the new H1N1 flu virus appeared a few weeks after Google's research results were published, enabling Google's system to assist the CDC by providing prompt and reliable tracking of the new outbreak.

The Internet of Things

Growing amounts of personal, medical, financial, professional, and business data are being collected through internet activities and stored in "the cloud"—vast collections of servers located on rural campuses operated by internet companies such as Amazon, Microsoft, Google, and Apple. The scale of data collection and storage will continue to expand exponentially with the growth of the "internet of things"—the sensors and associated software connected to the internet that are being embedded in personal devices, appliances, homes, cars, highways, machines, and workplaces. The result is a vast increase in scale of information, including personal information, collected and stored in the cloud and available for commercial exploitation.

The technology driving this expansion is the development of sensors that are smaller and cheaper, require less power, and have more computing capacity. These sensors will soon be everywhere, including the homes, cars, phones, watches, and fitness monitors of individuals, collecting detailed personal data including location, activities, health, and shopping preferences. This explosion of information will have many benefits, including more extensive health records, better traffic flows, safer cars, reduced energy use, more extensive monitoring of air and water quality, and more reliable machines, among many others. It will also lead to refinements in the relationship between retailer and customer, such as offering a special price to a customer in the vicinity of an item in a store based on the customer's present location and past buying profile. There is also potential for abuse, such as using health information in the cloud to adversely affect insurance decisions and rates and to identify desperate people and offer them high-interest payday loans.

Along with the benefits, this massive collection of data—and its exploitation and possible abuse for commercial purposes—could make a further contribution to a perception by many that the economy is not working for them, with consequent implications for governance. Security and privacy are two particular problems accompanying the increase in scale of data collected by ubiquitous sensors and stored in the cloud.

Security[18]

Security of the internet of things has been identified as a potentially serious issue. The sensors typically are small and have limited computing speed and power, limiting their ability to employ modern security techniques. Many have no ability to be upgraded to counter evolving security threats. Many are embedded in objects (such as appliances and utility meters) that have much longer lives (decades) than the lifetimes normally associated with high-tech equipment. They could still be deployed long after the company that created them no longer exists, with vulnerabilities to threats that emerge during their lifetimes. Many have little physical security, allowing attackers direct physical access. In many cases, the user is not aware of the device and does not monitor its status. The potential consequences of security problems with embedded devices could have implications for internet security generally; an unprotected refrigerator or television infected with malware could send harmful messages worldwide.

These considerations suggest that security should be a primary objective in the design and deployment of the internet of things. This could include the development of standards for best practices in design, security, and field upgradability of embedded devices. Industry standards, rather than government regulations, seem more practical for such a complex and rapidly changing environment.

Our systems are vulnerable. One of the first things we can do as we wrestle with this is deal with that reality and harden the systems we have. Then, design resilience into the new systems—not attempt to bolt it on after the fact.   —James O. Ellis, Jr.

Privacy[19]

The vast and exponentially growing amount of data collected from sensors that monitor individuals and their activities, including activities in their homes and cars, together with data collected and stored on their buying habits and professional activities, can be mined to produce detailed individual profiles for commercial purposes. The collection of this information may provide benefits for the device's owner, but often the supplier and manufacturer of the device benefit financially from the information as well. When data streams from multiple devices are combined and correlated, the result can be a quite invasive individual portrait. For example, internet companies can already accurately identify when an individual becomes pregnant, and the due date, from data on search queries and buying habits.

There is potential for conflicting interests between those whose personal data are exposed and those who collect this information and often sell it to third parties. Many of the "free" services on the internet are, in fact, paid for through the sale to brokers of personal data collected through tracking the activities and internet histories of individuals. From a commercial point of view, collection and sale of personal data can support a viable business model for many "free" internet applications. Some individuals, however, may see an unwelcome intrusion into a private space, often without consent or even awareness, adding to a perception that the economy is not working for them.

The legal situation with respect to privacy of electronic information is complex:

- Medical records must be protected; individuals have the right to see them and correct them. By law, medical information may be used or shared for certain specific reasons not directly related to an individual's care, such as monitoring quality of care or reporting disease outbreaks; individuals can generally learn who has seen their medical information. Otherwise, medical information cannot be used for purposes not directly related to care without permission. Vast amounts of health information are collected by entities not covered by existing privacy laws (from fitness devices and social media, for example). Conversely, achieving the full potential of medical science to detect, treat, and prevent diseases and to reduce costs could require evolution of the existing health information privacy regime.

- Financial institutions must notify their customers about their information-sharing practices and inform them of their right to opt out if they don't want their information shared with nonaffiliated third parties.
- Personal information on children under the age of thirteen cannot be collected without parental consent.
- Except for medical and financial information and information on children as outlined above, the privacy of personal data is governed only by the Federal Trade Commission Act's prohibition on "unfair or deceptive acts or practices." The FTC has taken action against firms (including Google and Facebook) for violating their published privacy policies, as a deceptive trade practice.[20] Otherwise, the collection and use of information, including personal information, is largely unregulated.
- By contrast, EU regulations require notice and consent for collection of personal data; the data can be used only for the stated purpose; and individuals may have access to their data and can make corrections.[21]

The United States is sleepwalking, whereas the Europeans and the Chinese have a strategy. The Chinese one is, "We need to control these network platforms. Let's have our own. Done." The European one is, "Well, we don't have any of our own because we're not good at technology. So let's just regulate the American ones." The US strategy is, "Everything is awesome." It's definitely not.   —Niall Ferguson

---

Inequality

The potential drawbacks of big data will not necessarily be evenly distributed. There are well-documented areas where data analytics systematically discriminate against the poor, including algorithms used for sentencing individuals, for screening job applicants, for setting rates for mortgages and automobile insurance, and for finding desperate people and targeting them with ads for high-interest payday loans.[22] In all these cases, nontransparent algorithms give low scores to the poor, in part because of where they live. Since both the algorithms and the data they use are not transparent, there is no way to learn the reasons for adverse decisions or to correct errors.

Advancing technology and data collection could allow the dynamic pricing now common for airline seats and sports events to spread to the rest of the economy, with prices set by nontransparent algorithms for goods and services offered to some that are different from the prices offered to others.

As algorithms operating on data that include detailed personal information come to play key roles in decision-making in the workplace, in retail, in the courtroom, and in insurance of all kinds, inequality and discrimination fostered by nontransparent algorithms can become a further source for a perception that the economy is not working for many citizens.

Part II. Mitigating the Impact of Advancing Technology

Transitions to New Jobs

Availability of New Jobs in New Occupations

While perhaps half of today's workers are in positions vulnerable to disruption by artificial intelligence, automation, and related technologies, other fields will expand and new jobs and new industries will be created. Machines will not generally replace humans for the foreseeable future. Many tasks will be best done by a combination of humans and machines, and many occupations requiring creative, management, and technical skills will remain human domains. The challenge will be to facilitate transitions of displaced workers to new occupations, including training in new skills.

A closely related issue concerns the large number of positions available today but not filled because employers cannot find qualified candidates. The Department of Labor estimates that in June 2017 there were 6.2 million job openings in the United States. "Employers struggle to fill well-paying jobs in health care, advanced manufacturing, information technology, construction, transportation and logistics with workers sufficiently skilled to handle the work."[23] Some of these existing unfilled job openings would potentially be available for displaced workers, with appropriate training.

It is therefore likely that jobs will be available for displaced workers, drawing on (1) the half of the workforce that will not be disrupted in the near term, (2) the new jobs created even as traditional jobs disappear, and (3) job openings that cannot now be filled due to lack of qualified workers.

New jobs would not necessarily be in the same locations, would not necessarily pay as well, at least initially, and would require investment in training to learn new skills, but they would eventually allow displaced workers to regain the considerable individual, family, and community benefits of gainful employment.

I think the respect for "skill" is something underemphasized in this country. . . . I remember Helmut Schmidt when he was German chancellor would take four, five, six of the top skilled tradespeople in Germany with him on his diplomatic trips abroad—you can imagine the respect those guys had when they came back. It was an intangible that we here need to really think about.　　　　　　　　　　　　　　　　　　　　　—Sam Nunn

Training for New Occupations: Partnerships between Employers and Community Colleges

A key resource that displaced workers can draw upon in moving to a second career is the widespread establishment of partnerships between employers and community colleges. Successful partnerships include:

- *A group of employers in an industry sector with similar workforce needs and located in the same geographic region (a portion of a state).* The group of employers defines the skills needed in the workforce. Participation of employers provides a concrete and visible prospect of a new job, which can serve as an incentive to devote the time, effort, and resources necessary to complete the retraining process. In some cases, employers can potentially offer internships as stepping-stones to new jobs.
- *A community college training program that provides the skills to match the workforce needs sought by the employers.* Partnerships between community college training programs and employers are established through industry participation on advisory boards, personal contacts between instructors and their industry counterparts, instructors with experience in industry, adjunct instructors currently in industry who teach courses at night, and feedback from graduates now working in industry. In this model, feedback from close partnerships with employers allows training programs to adapt to the evolving needs of employers (which can change rapidly). Experience has shown that in addition to technical skills, employers seek English and math skills, business skills (e.g., familiarity with Microsoft Office), and social skills (e.g., good communications, good interview). Many community colleges integrate the appropriate English and math into their technical programs so that students learn English and math in the context of their future careers.[24] Community colleges can also provide certificates and credentials required in certain fields. Moreover, community college

costs are low. In California, two-year training programs generally are in the $2,000 to $3,000 range, and much of that can be waived according to circumstances.

- *Tracking and feedback.* An important element of successful partnerships is the tracking of individuals after they complete the program. (Did they get a job? Did they get a job they trained for? What salary?) For successful career technical training programs, most of the individuals who complete the program get jobs in their field of study.[25] Tracking of employers is also important. (Did they interview individuals who completed the program? Did they hire those people? Did the applicants have the skills the employers were looking for?) These partnerships between employers and community colleges are dynamic entities, constantly evolving to meet the changing workforce needs of today's employers.
- *Other organizations can be involved to help overcome barriers faced by some job seekers.* Examples include organizations that provide day care support, transportation to and from training sites, and assistance in overcoming issues raised in background checks, including help in navigating the criminal justice system.

Connecting Displaced Workers to Available Jobs and Training

Even as jobs are eliminated by advancing technology, new jobs are created, and many jobs today are unfilled for lack of qualified workers. Community colleges in partnership with employers offer training programs at affordable cost in technical, business, and social skills that have successfully led to new jobs in new fields. The remaining question is how to connect displaced workers to new occupations with available jobs and to corresponding training programs.

In principle, an informed decision by a displaced worker would require information on new occupations with available jobs nearby, on the larger set of new occupations with available jobs at some distance from his or her current location, and the still larger set of new occupations with available jobs in other states, along with salary expectations and living costs for each case. Also important would be information on availability, cost, and time required for corresponding technical training necessary to prepare for a new job in a new occupation. The displaced worker could weigh options on a range of industry sectors, locations, salary structures, and training possibilities, and then make an informed decision on how to proceed, taking into account his or her personal and family situation.

Some resources are available to unemployed workers to help them acquire the large amount of information necessary to make an informed decision on a new occupation and perhaps a new location. Together with the states, the Department of Labor has established Job Centers, which provide online information on job openings and educational opportunities to support job seekers of all ages and circumstances, including a database of jobs that have been posted on the internet by employers and of training programs. Job seekers can explore information on a range of careers, prepare a self-assessment, learn about the job application process, write a résumé, and prepare for an interview. The 2,500 Job Centers throughout the country provide workstations that job seekers can use to access a database of jobs and training programs. The centers also offer counseling and sponsor job fairs with local employers.

While these resources available to job seekers represent a useful step in the process of transitioning to a new job, they are not a solution to the problem of finding new jobs for displaced workers:

- Not all jobs are posted on the internet. Many employers prefer candidates referred by current employees. Some jobs are reserved for internal candidates. Some employers work with recruiting agencies.
- Some jobs posted on the internet are not available for filling in the near term but are intended to collect information on candidates for potential future use.
- Not all training programs are included in the database.

- The large databases of jobs are generally accessed by a complex automated process using keywords, which is difficult to navigate. This problem is compounded by a lack of generally agreed-upon terminology for job titles. (For example, a "maintenance technician" could refer to a skilled person who controls the operation of an automated factory and is capable of trouble-shooting potential problems to keep it running smoothly. But the same words could refer to a janitor.)

A more fundamental problem is that the process of getting a job, especially a good job, is complex and difficult. Networking is generally a more productive approach than responding to postings on the internet. Many jobs go to candidates referred by current employees, who can be connected to job seekers by personal networks.

Social media plays an increasing role in job searches and hiring. Potential employers can learn a great deal about prospective employees from social media such as LinkedIn and Facebook. Social media can facilitate networking and can allow a job seeker to convey a résumé to a hiring manager. A résumé is likely to be effective only if it includes appropriate keywords that will be noticed by automated systems that scan résumés.

Gaining new skills, and a new job in a new field, with an adequate salary (probably less than the previous salary) is likely to be a difficult and time-consuming process. A good fit to a good job is likely to be the result of research, introspection, networking, and hard work. An appropriate strategy would begin with research into potential occupations with openings and community college training programs to acquire the necessary new skills. It would proceed with a mix of networking with friends, family, colleagues, and acquaintances, using social media, responding to internet job boards, and getting coaching on résumés and interviews.

There is a special opportunity here around land grant colleges, too. They have a tradition of application of knowledge to locally defined problems. And there is a correlation between land grants and the ability to sustain and even grow local manufacturing, especially when the land grant has that extension-service mentality of applied engineering.   —Ernest Moniz

Counselors, Coaches, and Caseworkers

Displaced workers would benefit from counselors or coaches who could assist them as they navigate this process to an acceptable outcome. Helping guide displaced workers to new careers is likely to be a task best undertaken by skilled humans working with automated systems, just as the combination of a doctor and a machine is the optimum approach to medicine and the combination of a teacher and a machine is the best approach to education.

Each displaced worker is a special case with his or her own abilities, ambitions, limitations, resources, and responsibilities. A good counselor would encourage reflection and self-assessment, help identify and assess options for new occupations, suggest networking possibilities, and recommend approaches to identifying and securing a job that have proved successful. A good counselor would be skilled in communication with a displaced worker (who would be going through a difficult and traumatic period in his or her life), skilled in the use of automated systems and other sources of information on occupations with available jobs and associated training programs, and aware of other resources and organizations that are available to help overcome obstacles that may arise in each case.

While skilled counselors, coaches, and caseworkers could be crucial to success in connecting displaced workers to available jobs and available training programs, such counseling is not currently an occupation supported by training programs. Upgrading the capabilities of counselors, coaches, and caseworkers may be the single most effective way to facilitate the transitions of displaced workers to new careers. The Markle Foundation, through its Skillful program, is seeking to develop a coaching corps by upgrading the skills of individuals now working as counselors at workforce centers and elsewhere, and by networking them together.

Another useful initiative would be to establish a system to track the progress of each unemployed worker through the process of transitioning to a new job. Collection of such data, while a difficult and expensive task, would lead to an understanding of what works and what doesn't and would help develop best practices for facilitating successful transitions.

Other Possibilities for Mitigation

While the great majority of new jobs will be in the private sector, public policies can play a key role in this process:

- *Economic growth.* A growing economy would make a substantial contribution to job creation, including jobs for those displaced by automation.
- *Lifelong education.* With the rapid advance of technology, education is no longer primarily for the young. Workers in all fields are well advised to stay current on technical and business skills and to learn new subjects in order to keep pace with developments in current occupations and provide a head start if and when it becomes necessary to prepare for a new career. Community colleges and internet courses provide low-cost education and training in a wide variety of fields.
- *Adjustment benefits.*[26] The prospect of further large-scale job dislocation has led to calls for consideration of a guaranteed basic income, which would be paid to everyone and provide a subsistence living without a job. Both individuals and society as a whole, however, would benefit from re-employing displaced workers in new occupations rather than paying them not to work. From the perspective of our society, there is no shortage of work to be done and problems to be solved. For individuals, income is not the only benefit of a job. A sense of self-worth and standing in the community are equally important. Rather than

paying people not to work, the focus should be on facilitating transitions to new jobs, with the possibility of support through the adjustment period. The narrowly defined Trade Adjustment Assistance program did little to counter the impact of job losses attributed to globalization. A broader adjustment benefit would cover layoffs due to automation and recession as well as foreign competition and would provide income and assistance for training and relocation during the transition period.

Studies were done in wartime Britain about how people reacted to the Blitz. People who felt most aggrieved, hurt, victimized weren't the people at the center of the blast radius. They were the people in the next ring out. Why? Because they could see, beyond them, people who weren't scarred at all.   —David M. Kennedy

---

Implications for Governance

While the coming disruption of the workplace by advancing technology promises to exceed past job losses attributed to globalization, many elements are in place that can support transitions to new occupations for affected workers. In particular, partnerships between employers and community colleges provide effective and affordable training for new careers. Governments can help mitigate job losses and their political consequences by providing sufficient support for community college technical training programs to meet the demand by displaced workers.

Also in place are programs to assist unemployed workers generally, which displaced workers can draw upon. Success in securing a new job in a new occupation with an adequate salary is likely, however, to depend heavily on individual efforts and networking skills. Governments could support the work of the nonprofit sector to upgrade the capabilities of coaches to help job seekers navigate the difficult path to a good fit with a good job in a new occupation.

Recognizing that the length of training programs for new careers often exceeds the six-month duration of unemployment benefits, governments could expand the Trade Adjustment Assistance program (which supports displaced workers in full-time training for up to two years) to cover job loss due to automation and recession as well as foreign competition.

Finally, governments could invest in tracking each unemployed worker through the process of training, job search, job placement, and subsequent career path. On the basis of such information, the process could be continually adapted to emphasize approaches that prove to work.

The goal here is to try to expand access to the positive aspects of these emerging technologies and, at the same time, protect citizens from their adverse consequences.   —Christopher Stubbs

---

Privacy

It would not be easy to develop practical solutions to privacy questions raised by the rapidly increasing collection of information from internet activities and from sensors that are already widely deployed and will soon be ubiquitous. Industry standards for privacy seem to be a more practical approach than regulations, given the complexity and rapidly changing nature of the problem.

Present practice is for companies engaged in internet commerce to post their privacy policies on their websites. Google, for example, has on its website a nine-page privacy policy with numerous links to added detail.[27] (As part of a 2011 settlement with the FTC on charges of violating its own privacy commitments, Google is required to implement a comprehensive privacy policy program with regular independent audits.) The privacy statement outlines the information Google collects and how this information is used and shared. There is a detailed description of the information automatically collected and stored on server logs every time a user makes a request or views content on Google sites.

Automated Google systems analyze user activities to support, among other things, tailored advertising. Google offers options for opting out of storage and sharing of certain types of information.

Other internet companies post privacy policies that are similar in size and complexity, but differ in detail. Companies are generally open about the vast amount of information collected and how it is used and shared. The commercial use and sale of this information support "free" internet services. As noted earlier, the FTC can and does take action against companies for practices inconsistent with their stated privacy policies.

In addition to these privacy policies of individual companies, there are collective self-regulatory efforts:

- The Network Advertising Initiative, a consortium of online advertising providers, has produced a code of conduct for data collection, transfer, and use practices in online advertising, including an opt-out mechanism that individuals can exercise on its website.[28] These are the companies that provide the infrastructure behind interest-based advertising, and most internet ads displayed in the United States involve one or more subscribers to the code. The FTC has taken enforcement actions against companies that failed to honor opt-outs.[29]
- Internet companies involved in K-12 education services, including Apple and Google, have taken a pledge not to collect, maintain, use, or share student personal information beyond that needed for authorized educational purposes.[30]

The present framework of published voluntary privacy policies on data collection, use, and distribution, including provision for opting out of certain uses, backed up by the potential for FTC enforcement in event of violation of those stated policies, can be the baseline for consideration of further steps:

- Greater transparency for data brokers.[31] Several nearly invisible entities called data brokers buy and sell personal information collected from numerous sources, including internet activities (web pages visited, items purchased, social media posts, etc.), largely without consumer knowledge. They use it to form detailed descriptions of the online and offline lives of nearly everyone.[32] These files on nearly every individual include health and financial information. This personal information is used to facilitate targeting of advertising (via web pages, email, and direct mail), to help identify fraudulent transactions, and to locate people. There is now no meaningful way for individuals to know the full spectrum of information collected about them and the way it is reused. The problem will continue to grow as the data collected by ubiquitous sensors expands the information collected on personal activities. Following the lead of the online advertising industry, the data services industry could collectively produce a website identifying data services companies, describing their data practices, and providing a mechanism for individuals to gain access to their data and to opt out of certain uses.

- Consumer choice for unrelated use of data.[33] Individuals could also have a choice about uses of information about them that are not related to the service provided. Many internet applications collect data not related to the service performed, which can then be sold. For example, some applications that enable a smartphone to be used as a flashlight collect GPS data on the location of the user. Geolocation information is obviously not necessary for the flashlight; the tracking data are collected and sold to provide a separate income stream. This collection of extraneous data is disclosed, and while these disclosure statements are rarely read before downloading an app, analysis of disclosure statements has led to public awareness of the risks of flashlight apps.[34] Industry standards could require that collection of personal data not related to the service provided be prominently displayed in disclosure statements so that consumers could make an informed choice.

Enhancing Public Information on Privacy Policies and Algorithms

A great deal of public information is already provided in lengthy, complex notifications of privacy policies that are often clicked but rarely read. A nonprofit organization could usefully be established whose mission is to carefully read and compare privacy policies and then publish reports calling public attention to privacy risks, comparable to the reports that brought public attention to the risks of phone flashlight applications.

Stronger protections, such as a requirement for consent to data collection, do not seem realistic given the deployment of sensors already under way and the broad acceptance of collection and use of electronic data in many areas, such as the workplace, banking, and retailing.

Most Americans seem to accept the current provision of "free" services supported by the sharing of personal information, even as they may not be fully aware of the scope and detail of the personal information involved. Individuals who have strong feelings about the privacy of their personal electronic data could take appropriate action based on existing public disclosures of privacy policies, perhaps augmented by reports from an organization dedicated to advising the public on privacy risks of personal data collection and use, and accept whatever inconvenience is associated with that decision.

The European Union is moving in the opposite direction, placing increasingly stringent restrictions on the collection and sharing of personal data. Given the low level of expressed concern about privacy of electronic personal information in the United States, establishment of an organization to increase awareness of privacy risks based on publicly disclosed privacy policies seems a more practical approach for this country.

While there would be resistance to exposing proprietary business practices, some degree of transparency could also be established for algorithms that are used in decisions that increasingly affect the personal and professional lives of many citizens (including algorithms used for sentencing individuals, for screening job applicants, for determining mortgage rates, and for setting rates for insurance of all kinds). This is another area where consideration could be given to the establishment of a nonprofit organization staffed with appropriate expertise that could assess algorithms widely used in the courtroom, in the workplace, and in commerce, and report to the public on the capabilities and limitations of these algorithms, including the accuracy of their predictions.

We have these frightening technologies. What do we do with them? I want to express the sentiment that technology will always go forward. It cannot be stopped. Anything that thinks about stopping it or even slowing it down is just not going to work, and so we have to deal with it. —Persis Drell

---

2

TECHNOLOGICAL CHANGE AND THE FOURTH INDUSTRIAL REVOLUTION

T. X. Hammes

Klaus Schwab, executive chairman of the World Economic Forum, coined the term "fourth industrial revolution" for "the staggering confluence of emerging technology breakthroughs, covering wide-ranging fields such as artificial intelligence (AI), robotics, the internet of things (IoT), autonomous vehicles, 3D printing, nanotechnology, biotechnology, materials science, energy storage and quantum computing, to name a few."[1]

This chapter argues that the "staggering confluence" will drive deglobalization, diffuse military power to smaller entities, and change the character of competition between the great powers. If managed well, these advances will provide significant advantages to the United States in its competition with both China and Russia, even as it allows smaller powers to challenge it. If managed poorly, the convergence of technology will severely undercut US security.

Deglobalization Has Started

The Economist defines globalization as the "global integration of the movement of goods, capital and jobs."[2] Over the last seven decades, the combination of lower labor costs, increasingly efficient freight systems, and trade agreements based on Bretton Woods provided major regional cost advantages for manufacturing. The resultant globalization transformed agricultural societies into industrial powerhouses.

While the process may seem irreversible, the globe has already gone through repeated periods of deglobalization, ranging from the decades that followed the First World War and the 1970s oil shocks back to the centuries that followed the collapse of the Roman Empire. As recently as 2008–09, a global financial crisis once again devastated global trade. Academics, think tanks, and respected news journals all speculated that the collapse of the financial markets would result in a rapid and long-term decrease in global trade. In February 2009 *The Economist* noted that each element of globalization appeared to be in trouble.[3]

In fact, the global economy defied deglobalization predictions and bounced back. World Bank statistics show global merchandise trade, as a percentage of GDP, recovered quickly from the 2009 crisis, almost reaching precrisis levels by 2011. Optimism returned that global trade would once again lead global economic growth.

The optimism was premature. Even as worries about deglobalization faded, trade, as a percentage of global GDP, flattened and then declined. While the world economy continued to grow slowly from 2011 to 2016, the growth of international trade lagged behind—instead of exceeding—the growth rates. Most alarming, the steepest decline was from 2014 to 2015, when the rate of manufacturing decline approached that of 2008–09 even though the global economy grew by 0.3 percent (figure 2.1).[4]

It was not just manufacturing. When the World Bank added services to merchandise, total trade followed the same pattern. Growth, as a percentage of global GDP, declined slightly each year since 2012 (figure 2.2).[5] Once again, US and Chinese trade statistics moved in parallel.

In summary, according to the World Bank the total value of global merchandise trade increased sharply from 1995 to 2008, dropped sharply in 2008, climbed steeply until 2011, then flattened from 2012 to 2014 and fell sharply in 2015—dropping from $18.995 trillion to $16.482 trillion.[6] Thus, two of *The Economist*'s key measures of globalization—

FIGURE 2.1   Merchandise Trade as a Share of GDP, 2008–15

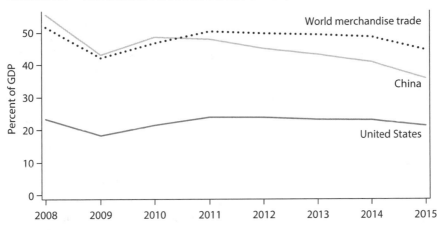

Source: World Bank, "Merchandise Trade (% of GDP)," accessed December 28, 2016, http://data.worldbank.org/indicator/TG.VAL.TOTL.GD.ZS/countries?display=graph.

FIGURE 2.2   Total Trade (Manufacturing and Services) as a Share of GDP, 2008–15

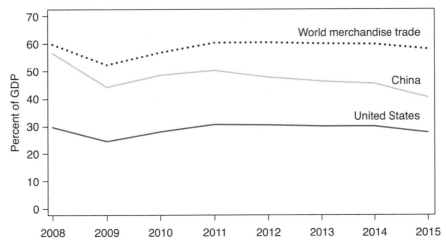

Source: World Bank, "Trade (% of GDP)," accessed December 28, 2016, http://data.worldbank.org/indicator/NE.TRD.GNFS.ZS/countries/1W-CN-US?display=graph.

the movements of goods and of services—have in fact been declining since 2011.

As one would expect, another measure of international trade activity, global financial flows, declined very sharply during the crisis. However, unlike global trade flows, financial flows never recovered (figure 2.3). As noted in a policy brief for the Centre d'Études Prospectives et d'Informations Internationales in Paris, "The 'Great Retrenchment' that took place during the crisis has proved very persistent, and world financial flows are now down to less than half their pre-crisis levels."[7]

These decreases in financial flows hit developing countries particularly hard. Capital inflows to these countries initially rebounded after the 2008 crisis but slowed again after 2010 and then turned negative in 2014. In 2015, over $700 billion in capital left developing economies, greatly exceeding even the $145 billion net outflows during the Great

FIGURE 2.3   Global Gross Financial Flows, 2005–14

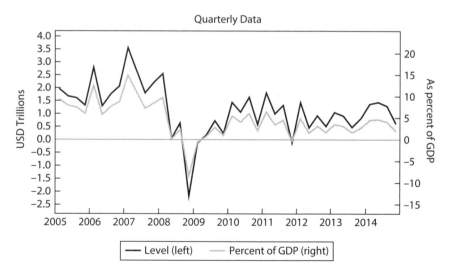

Source: Matthieu Bussiere, Julia Schmidt, and Natacha Valla, "International Financial Flows in the New Normal: Key Patterns (and Why We Should Care)," CEPII (Centre d'Études Prospectives et d'Informations Internationales), March 2016, accessed May 26, 2016, www.cepii.fr/PDF_PUB/pb/2016/pb2016-10.pdf.

FIGURE 2.4   Net Financial Flows to Developing Economies, 2006–15

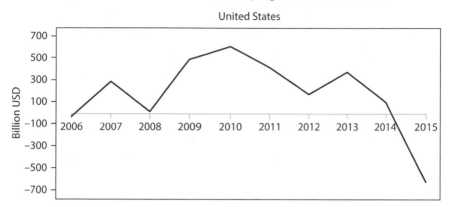

Source: United Nations, 2016, "World Economic Situation and Prospects," http://www.un
.org/en/development/desa/policy/wesp/wesp_current/2016wesp_ch 3_en.pdf.

Recession (figure 2.4).[8] In contrast, foreign direct investment (FDI) into
the United States is growing rapidly. In 2016, FDI flows into the United
States reached $391 billion, more than double the $171 billion inflow of
2014. Outflows in 2016 were only $299 billion.[9] Thus, in 2016, the
United States saw a net inflow of investment capital of $192 billion. In
2015, as shown by the latest statistics available from the Department of
Commerce, nearly 70 percent of the FDI was invested in the manufacturing sector.[10]

Amplifying the effect of reduced financial flows has been their region-
alization or balkanization. After decades of coming closer together,
global markets and banking systems are pulling apart. While cross-
border goods, services, and financial flows represented 53 percent
of the world economy in 2007, they are a mere 39 percent now.[11] In
November 2014, Kristin Forbes, a leading economist and member of
the Bank of England's external monetary policy committee, noted that
gross cross-border capital flows for advanced economies had dropped
to levels not seen since 1983. "Financial globalization has sharply
reversed and shows little signs of returning even to levels of the late
1990s," she said, urging a rethink of the "assumption that global financial

integration is an unstoppable trend."[12] On October 5, 2016, the *Wall Street Journal* noted the slowdown is not limited to developed economies but is also hitting emerging economies in both goods and services.[13]

In March 2016, the *Harvard Business Review* counterargued that globalization was not slowing but still increasing rapidly. It stated, "Cross-border data flows have grown by a factor of 45 over the past decade, and they're projected to post another ninefold increase by 2020."[14] McKinsey Global Institute analysts wrote that data flows accounted for $2.8 trillion of value—exerting a larger economic impact than the global trade in physical goods.[15]

While an increase from near zero to $2.8 trillion is very impressive growth, it should be kept in perspective. According to the World Bank, global GDP was $74 trillion in 2015.[16] Further, the authors of the study admit it is very tricky to translate the number of terabytes of data flowing across borders to a dollar value. For our purposes, the key issue is whether the data flows are in fact increasing globalization—and that is impossible to tell. Globally, 70 percent of internet traffic during peak hours in 2015 came from video and music streaming; this is expected to increase to 82 percent by 2020. According to Cisco, "A growing number of M2M (machine to machine) applications, such as smart meters, video surveillance, healthcare monitoring, transportation, and package or asset tracking, are contributing in a major way to the growth of devices and connections. . . . By 2020 the consumer share of the total devices, including both fixed and mobile devices, will be 74 percent, with business claiming the remaining 26 percent."[17] It is difficult to see how data that are heavily about entertainment and day-to-day operation of personal devices contribute heavily to globalization. This subject certainly requires further research.

Factors Driving Deglobalization

No single industry or social development is driving deglobalization. It is being driven by the cumulative effect of technological, political, and social trends across the globe. Perhaps the primary driver, least subject to reversal, is the elimination of regional labor cost advantages that encouraged manufacturers to locate their production in low-labor-cost regions. Robotics, 3D printing, and artificial intelligence are driving manufacturers to reconsider not only how and what they make, but where they make it.

As I look at innovations, in a sense they are lacking. Our produc-tivity growth is actually very, very low now in the United States and other countries. So I think there's a danger trying to stomp out these things when they may be what are actually going to make people's lives better.   —John B. Taylor

---

How We Make Things

The movement to automation is a global trend that is accelerating. According to Boston Consulting Group, 25 percent of all manufacturing will be automated by 2025. In 2015 electronics manufacturing robots' operating costs were the same as Chinese labor, about $4 per hour.[18] Since then, robot costs have continued to decrease, but Chinese labor costs are increasing rapidly.[19]

The world is on the very front end of the shift from labor to automation. A new Price Waterhouse Cooper study estimates 38 percent of US jobs could be lost to automation in the next fifteen years.[20] Robot sales are expected to reach four hundred thousand annually by 2018.[21] This estimate does not account for the newly developed "cobots," or collaborative robots. They assist human workers and thus dramatically increase human productivity. At an average cost of only $24,000, they will appeal strongly to the smaller companies that account for 70 percent of global manufacturing.[22] Further, converting to automation creates a virtuous circle. A Price Waterhouse Cooper survey showed 94 percent of those CEOs who had already adopted robots say the robots increased productivity.[23] Thus, those who buy robots are encouraging others to buy them.

Even as robots are changing traditional manufacturing, 3D printing, also known as additive manufacturing, is creating entirely new ways to manufacture a rapidly expanding range of products. The ability to print everything from medical devices to aircraft parts to buildings and bridges, combined with a recent order-of-magnitude increase in the speed of printing, is already challenging traditional manufacturing. In April 2016, Carbon Inc. introduced a commercial 3D printer that is one hundred times faster than previous printers.[24] It plans to push the speed to one thousand times faster. The Department of Energy's Oak Ridge National Laboratory is partnering with Cincinnati Inc., a manufacturer of high-quality machine tools, to develop a process to print metal two hundred to five hundred times faster.[25] In May 2017, MIT developed rapid liquid printing, which is faster and also allows a much greater range of products.[26]

Commercial firms are exploiting these advances. United Parcel Service has established a new facility called Direct Digital Manufacturing with one hundred printers that will all run twenty-four hours a day, seven days a week, and require just one employee per eight-hour shift.[27] UPS plans to expand the plant to a thousand printers and is already establishing additional 3D printing facilities worldwide.[28] In a Price Waterhouse Cooper survey, 52 percent of the CEOs expect 3D printing to be used for high-volume production in the next three to five years. This is up from 38 percent only two years ago.[29]

What We Will Make

Three-dimensional printing will have a major impact on manufacturing by bringing two other changes: design for purpose and mass customization. For the first time, designers can design an object to optimally fulfill its purpose. Current manufacturing techniques often require that optimal design be subordinated to manufacturing limitations. While the designer may have envisioned the most efficient form for a product, that form may be impossible to machine or build. Three-dimensional printing frees the designer to create virtually any form and see it printed to specification. Design for purpose is already changing how we make things and will have a major impact on production.

The second impact will be mass customization. With some forms of advanced manufacturing, 3D printing in particular, there is essentially no cost to changing the specifications of the object being printed. Thus, every item can be customized. Globally, manufacturers are seeking to respond to consumer demand for unique items. Only local facilities using advanced manufacturing techniques can deal with the rising demand for mass customization in everything from clothes to cars. This is driving manufacturers to return manufacturing to the markets they serve.

Where We Make Things

By reducing labor costs while simultaneously increasing customization, productivity, and quality, these new technologies are bringing manufacturing back to America. The United States lost manufacturing jobs every year from 1998 to 2009—a total of eight million jobs. But in the last six years, it regained about one million of them.[30] With the cost of labor no longer a significant advantage, it makes little sense to manufacture components in Southeast Asia, assemble them in China, and then ship them to the rest of the world when the same item can either be manufactured by robots or printed where it will be used.

Another major factor accelerating the shift of manufacturing back to the United States is the reduction in risk to intellectual property. Local manufacturing also reduces shipping costs and reduces—even, in some cases, eliminates—inventory. "Just in time" local production means no finished items in stock—only a small supply of input materials.

Service Industries Are Coming Home Too

Service industries are following suit as artificial intelligence takes over more high-order tasks. Call centers are already moving from low-wage areas back to the United States. Early adopters of AI-driven customer service centers like United Services Automobile Association have achieved very positive results.[31] Pairing AI with humans has resulted in lower costs (fewer humans) and higher customer satisfaction as the problems that AI can't resolve are handled by Americans.

Nor is artificial intelligence limited to routine call center tasks. The sophistication of artificial intelligence is growing so quickly that in 2015 the Georgia Institute of Technology employed a software program it named Jill Watson as a teaching assistant for an online course. All of the students thought Ms. Watson was a very effective and helpful teaching assistant. None guessed she wasn't human.[32] Baker & Hostetler, a law firm, announced it has hired her "brother," Ross, also based on Watson, as a lawyer for its bankruptcy practice.[33]

Even as AI moves into sophisticated tasks, robotics will also take over mundane tasks like delivery, stocking, cleaning, etc. Many back office tasks will also go the way of telephone operators and call center staffs. Service jobs that require the human touch or have to deal with non-routine tasks will remain, but massive numbers of humans will be replaced. Further up the knowledge ladder, artificial intelligence is already handling tasks formerly assigned to associate lawyers, new accountants, new reporters, new radiologists, and many other special-ties. In short, nonroutine tasks, whether manual or cognitive, will still be done by humans while routine tasks—even cognitive ones—will be done by machines. And this is not a new phenomenon. Computer tech-nology has been eating jobs since 1990 (figure 2.5).[34] A study by the St. Louis Federal Reserve indicates that if routine jobs had grown at the same rate as nonroutine, America would have fifty million more jobs.[35] The jobs were lost to automation.

With labor costs much less of an issue, better communications links, better infrastructure, more attractive business conditions, and effective intellectual property enforcement will encourage services to return to developed nations. The few, more complex questions that require human operators will be better handled by native language speakers who are intimately familiar with the culture.

FIGURE 2.5   Routine vs. Nonroutine Jobs and Cognitive vs. Manual Jobs, 1983–2016

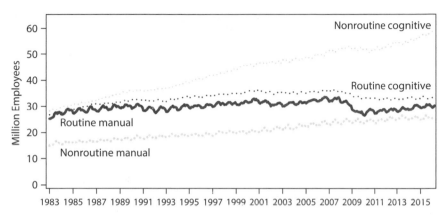

Source: US Bureau of Labor Statistics, 2017, "Current Population Survey (Household Survey): Series LNU02032201, LNU02032204, LNU02032205, LNU2032210, LNU2032211, LNU2032212." As suggested by Maximiliano Dvorkin, "Jobs Involving Routine Tasks Aren't Growing," Federal Reserve Bank of St. Louis, https://www.stlouisfed.org/on-the-economy/2016/january/jobs-involving-routine-tasks-arent-growing, and "Job Polarization," The FRED Blog, St. Louis Fed, https://fredblog.stlouisfed.org/2016/04/job-polarization.

I like the example of freestyle chess, where the human provides the strategy and a machine explores tactics and potential moves. A good player plus a machine can beat any human and any machine.
　　　　　　　　　　　　　　　　　　　—James Timbie

Other Technologically Driven Trends

The reduced demand for transportation fuel, alternative energy technologies, and increased energy efficiency are already reducing the global movement of coal and oil. At the same time, wind, solar, and thermal are growing rapidly. In 2014, 58.5 percent of all net additions to global power systems were renewables.[36] In 2015, 68 percent of the newly installed capacity in the United States was renewable.[37] Energy from renewable sources can be traded across adjacent borders, but not globally. Thus the increases in renewable energy production will steadily reduce the long-range shipment of coal, oil, and gas.

With Morgan Stanley predicting that more electric vehicles will be sold in 2040 than gas vehicles, the source of transportation fuel will move from petroleum to electric energy.[38] Fracking, alternative energy, and new efficiencies have already dramatically reduced the US need for imported energy.[39] If other nations can make similar advances in these areas, large segments of the energy market will become local. Growth in these energy sources will slow and perhaps eventually reverse the global trade in gas and oil.

In 2030, the United States will be able to produce about 95 million metric tons of liquid natural gas, which is basically larger than any other place on the earth, including the Middle East. So having that ability to interface with other countries using the energy resource, I think, could be a very big part of how the United States thinks of global interdependence. —Tsunehiko Yanagihara

---

A further driver of fragmentation is the effort by authoritarian governments to segment the internet. In his book *Splinternet,* Scott Malcomson notes that when the web took off in 1995, Russia and China started saying they wanted "more control over our particular parts of it."[40] They have been trying to gain that control ever since. Initially, total control was considered an impossible goal, but China has steadily improved its ability to control what people can access inside its territory. As Simon Denyer described recently in the *Washington Post,* "What China calls the 'Golden Shield' is a giant mechanism of censorship and surveillance that blocks tens of thousands of websites deemed inimical to the Communist Party's narrative and control, including Facebook, YouTube, Twitter, and Instagram."[41] There is to be no turning back. In 2017, China passed a new law on cybersecurity that codified and strengthened its control of the internet to include requiring corporate data to be stored inside China.[42] In actively discouraging outside information companies, China and other totalitarian nations have decided that connectivity's threats to stability exceed the benefits of global connection. Restricted access to the internet will inevitably reduce these nations' participation in the global economy.

During the 2016 US elections, both candidates reflected the changing attitude among Americans concerning trade. While the assumption that global trade is good may still exist among policy-makers and economists, it is rapidly fading among voters. In 2002, the Pew Research Center found that 78 percent of Americans supported global trade. By 2008, the percentage had fallen to 53 percent.[43] In 2014, when Pew changed the question from whether trade was good for the nation to whether trade improved the livelihood of Americans, favorable ratings plunged. Only 17 percent of Americans thought trade led to higher wages, and only 20 percent believed it created new jobs.[44]

The public mood has begun to change in Europe too. Dissatisfaction with past policies and priorities pushed by elites has created a palpable populist narrative. Voters have focused inwardly on economic and cultural issues. Nationalist parties have been present in various nations in Europe for decades, but a confluence of factors has raised their profile and power dramatically. Brexit and the strengthening of right-wing parties across Europe from Poland to Denmark seem to reflect a political pushback against globalization. While the Macron victory in France provided relief, it is important to remember that the far right still made major advances in France. The National Front's success was repeated in Germany's September 2017 election where the hard-right Alternative for Germany surged to finish third and became the first far-right party to win seats in the Bundestag since 1961.[45]

Since the financial crisis of 2008, more than 3,500 protectionist measures have been instituted globally, as well as numerous additional administrative requirements that have increased the difficulty of conducting international trade.[46] The trend toward protectionism is not slowing. On November 11, 2016, the *Wall Street Journal* reported that the "Group of 20 largest economies imposed a total of 85 new measures that restrict trade between mid-May and mid-October."[47] The December 2016 meeting of the World Trade Organization's Trade Policy Review Body noted that of the 2,978 measures recorded since 2009, only 740 had been removed.[48]

In October 2016, the *Wall Street Journal* noted, "The risk now is that politics, economics and finance are combining in a way that threatens to throw globalization into reverse, hanging a sword of Damocles over the world economy. Years of sluggish growth, stagnant wages and rising inequality are fueling a growing political backlash against what some say is unfair competition from foreign firms and foreign workers across developed countries, most clearly evident in the US election and in the Brexit vote."[49]

This shift from globalism has not been limited to the United States and Europe. In its 2016 year-end roundup issue, *The Economist* noted:

> Indeed for most people on Earth there has never been a better time to be alive. Large parts of the West, however, do not see it that way. For them progress happens mainly to other people. Wealth does not spread itself, new technologies destroy jobs that never come back, an underclass is beyond help or redemption, and other cultures pose a threat—sometimes a violent one.[50]

Social Reactions

The American public and much of Europe have turned against globalization. Three of the primary drivers are unemployment, rising inequality in wealth, and concern for the environment. In fact, new technologies have historically caused significant political and social disruption when they shifted the basis for wealth generation. If the historical pattern holds, the fourth industrial revolution will cause major disruptions in employment as the knowledge and skills required change significantly. For those who can adapt or who possess the capital to invest in the new technology, the future looks good. As we have seen in previous industrial revolutions and over the last thirty years in the United States, it has increased the income and wealth inequity greatly. The negative impact has been—and will continue to be—focused on the older, less educated, and poor. This will create much greater demands on US social safety networks.

The third driver, the environment in the form of global warming, has resulted in a great deal of political organization on both sides of the issue. There is a growing "buy local" movement driven by the desire to reduce the environmental impact of production. Local manufacturing, services, and food production create jobs near the consumer while dramatically reducing transportation energy and packaging waste. The people who support these movements see continued global engagement as a negative element in community life.

We're in a very, very pivotal time right now, and creating a literate, scientifically literate public is essential for the future of the country and the health of our globe.   —Lucy Shapiro

---

Will Deglobalization Continue?

The key question is whether recent reductions in trade and financial flows are simply a cyclical downturn or are actually signs of a sustained long-term shift in global trade patterns. I argue that the convergence of new technologies, reinforced by political and social forces, will dramatically change what, how, and where we make things. Trends in energy production, politics, and the internet's balkanization will accelerate these changes, retarding—if not entirely reversing—globalization. Over the next decade or two, these trends will result in the localization of manufacturing, services, and energy production. Profits, politics, and social trends will all drive trade patterns toward regional trade blocks, thus reducing global networks and fundamentally altering the international security environment. Even if political and social movements were to shift back in favor of globalization, the converging technologies will still make it more profitable to produce in the target market, and that will fundamentally change the global economy.

World orders are not self-sustaining. They can and do fail in very dramatic and impactful ways. Economic interdependence among nations is not in and of itself enough to prevent the dissolution of world orders or indeed world wars.   —James O. Ellis, Jr.

The Fourth Industrial Revolution Is Changing the Character of War

The fourth industrial revolution will drive massive changes in the economic, political, and social spheres and will inevitably change warfare too. Clausewitz's primary trinity of passion, chance, and reason will continue to define the fundamental nature of war and technology will not eliminate the fog and friction. However, fourth industrial revolution technologies are already changing the character of war—and these changes are accelerating.

Twenty-First Century Technologies

As just a sampling, the following will focus on how the dramatic improvements in nano-energetics, artificial intelligence, drones, and 3D printing are producing a revolution of small, smart, and cheap weapons that will redefine the battlefield.

Nanotechnology

The field of nanotechnology will have major impacts across societies and conflicts. One notable application has been nano-explosives. As early as 2002, nano-explosives generated twice the power of conventional explosives.[51] By 2014, open-source literature claimed nano-aluminum created ultrahigh burn rates, which give nano-explosives four to ten times the power of TNT.[52] The obvious result is that small platforms will carry great destructive power.

Artificial Intelligence

Two areas of artificial intelligence are of particular importance in the evolution of small, smart, and cheap weapons: navigation and target identification. They are essential for autonomy, and autonomy will be required if these weapons are to be employed in the thousands. The global positioning system (GPS) has proved satisfactory for basic autonomous drone applications such as the Marine Corps's K-MAX helo-drone in Afghanistan.[53] However, GPS will be insufficient for operations in narrow outdoor or indoor environments, dense urban areas, and areas where GPS is jammed. Academic and commercial institutions are working hard to overcome the limitations of GPS to provide truly autonomous navigation for drones.[54] The University of Pennsylvania has already developed a quadcopter that "uses a smartphone for autonomous flight, employing only on-board hardware and vision algorithms—no GPS involved."[55] Essentially, this recreates the inertial plus visual navigation system that guided the Tomahawk cruise missile before the advent of GPS.

While such a system cannot be jammed, it would only serve to get a drone to the target area but not to ensure it could hit a specific target. At that point, the optical systems guided by AI could use onboard multispectral imaging to find a target and guide the weapon.[56] In short, the AI necessary for many types of autonomous drone operations currently exists—and is operating aboard small, commercial drones.

It is exactly that autonomy that makes the technological convergence a threat today. Because such drones will require no external input other than the signatures of the designated target, they will not be vulnerable to jamming. Not requiring human intervention, the autonomous platforms will also be able to operate in very large numbers. They can be launched as precision strike weapons or sent to an area and then commence active hunting for prioritized targets. They can also be programmed to wait patiently prior to launch or even proceed to the area of the target and then hide until a specified time or a specified target is identified.

Drones

Clearly, commercial drone capabilities have increased dramatically in the last five years, and usage has spread widely. Still, small drones can only carry a limited payload. This limitation can be overcome with two separate approaches. First is the use of explosively formed penetrators (EFPs).[57] The second, and less technically challenging, approach is to think in terms of "bringing the detonator."

Weighing as little as a pound, EFPs can destroy even well-armored vehicles. In Iraq, coalition forces found EFPs in a wide variety of sizes—some powerful enough to destroy an Abrams tank. Others were small enough to fit in the hand—or on a small drone.[58] And of course nano-explosives could dramatically increase the destructive power of the weapons. The natural marriage of improvised explosive devices (IEDs) to inexpensive, autonomous drones is virtually inevitable. New 3D printers can manage the previously very difficult task of forming the copper plates that become the projectiles.[59] Thus, we can expect small and medium-size drones to pack a significant punch even against protected targets.

The "bring the detonator" approach applies to aircraft, vehicles, fuel, chemical facilities, power distribution networks, and ammo storage facilities as targets. In each case, the objective is to use the drone to detonate the large supply of explosive material provided by the target. Even a few ounces of explosives delivered directly to the target can initiate the secondary explosion that will destroy it.

Even as small commercial drone sales move into the millions, larger commercial versions are showing capabilities normally associated with much more expensive, manned platforms. Today, the Aerovel Flexrotor has a range of two thousand miles and forty-plus hours of endurance. It flies autonomously, launches and recovers vertically, and has a suite of onboard multispectral sensors.[60] It is currently used for surveillance and survey of large remote areas at a fraction of the cost of manned plat-forms. Defiant Labs' DX-3 is designed specifically to conduct geographic and resource surveys of remote regions to include Canada's environmen-tally hostile far north. It launches from a truck, navigates autonomously for up to nine hundred miles, and carries a variety of payloads.[61]

Military drones are also pushing performance parameters. The decade-old Israeli-designed Harpy and the US Navy's Tern are vertical launch, autonomous drones that carry sixty- and five-hundred-pound payloads respectively over one thousand kilometers. Kratos's XQ–222 Valkyrie can deliver five hundred pounds out to three thousand miles.

Additive Manufacturing

This brings us to how additive manufacturing/3D printing will allow drones to be employed as rounds of ammunition. Researchers at the University of Virginia 3D-printed and assembled an autonomous drone with a range of twenty kilometers (forty if sent one way) for only $800—most of the cost was the Android phone used for navigation. It took about a day to print the body.[62] Thus a small factory with only one hun-dred 3D printers using the new Carbon 3D printing technology could produce ten thousand such autonomous drones a day. If the UPS fac-tory of one thousand printers becomes the standard, it could produce one hundred thousand a day. The key enabler will be Department of Defense preparation so that it can provide the unique elements such as explosives, guidance, and targeting software.

It is one thing to have access to thousands of drones. It is quite another to have the logistics and manpower available to effectively employ them. One method that demonstrates it can be done is the Chinese system that mounts eighteen Harpy unmanned combat air vehicles (UCAVs) on a single five-ton truck.[63] The Chinese can transport, erect, and fire these fairly large drones (nine-foot wingspan) with a two-person crew. A similar-size truck could be configured to carry hundreds of US

Switchblade-size or Israeli Hero-size drones.[64] Thus a single battery of ten trucks could launch thousands of autonomous, active hunters over the battlefield. The key is autonomy, since it would be nearly impossible to provide sufficient pilots for each battery.

Drones could also be configured in standard twenty-foot shipping containers, which would create two major advantages. First, any truck or ship that can carry a container becomes a potential weapons platform. Second, it is almost impossible to preempt weapons in this configuration—there are simply too many twenty-foot containers to be targeted in a preemption campaign. The Russians have noted this and already sell their powerful Club K cruise missile system in a container indistinguishable from a standard shipping container.[65]

Implications for the Modern Battlefield: Land, Sea, Air, Space, and Cyber

Drone swarms may make defense the dominant form of warfare in ground, air, sea, and space domains and be able to attack the physical elements of the cyberdomain. As noted, commercially available autonomous drones have ranges out to two thousand miles and payload up to dozens of pounds. Military drones under development have ranges out to three thousand miles with payloads up to five hundred pounds. Combining nano-explosives, artificial intelligence, and additive manufacturing could create units capable of launching hundreds to tens of thousands of smart drones in wave attacks at ranges exceeding today's ground fire support systems or tactical aircraft. The cumulative impact of improved missiles and autonomous drones in large numbers will be to make domain denial much easier than domain usage.

Land Domain

On land, the family of small, smart, and cheap autonomous weapons may create a situation similar to that between 1863 and 1917 when any person in range moving above the surface of the ground could be cheaply targeted and killed. The result was static trench warfare. Like that historical period, the defense dominated because defenders could dig in. The offensive force had to expose itself to move. Today, missile and drone launchers can be protected in bunkers or simply by blending in to the congested terrain ashore. They can be inside any number of buildings or shelters or simply dug in. They will not emit any signal until they move to fire. In contrast, a system on the offensive will have to move in order to cover the distance from its home base to the objective. It will both create a targetable signature and have only the protection it can carry. How will a mechanized brigade be able to move if an enemy can launch 10,000 autonomous drones to hunt and kill its vehicles?

Sea Domain

Obviously, swarms of autonomous drones can also threaten any naval force trying to project power ashore. The drones will not need to sink a ship to achieve a mission kill. For instance, a drone detonating against an aircraft on the deck of a carrier or firing a fragmentation charge against an Aegis combat system's phased array radar can degrade that platform's capabilities to the point of a mission kill.

Undersea weapons will provide a much greater challenge to navies. Vietnam, Japan, South Korea, Australia, China, and Indonesia are all upgrading their submarine forces. However, a submarine force is expensive, complex, and difficult to operate. Unmanned underwater vehicles may provide a much cheaper deterrent. Since 2014, the US Navy has operated a fleet of semiautonomous underwater gliders that have the capability to operate for five years without refueling by changing their buoyancy.[66] They can patrol for weeks following initial instructions, then surface periodically to report and receive new instructions. Similar drones are being purchased globally for about $100,000 each, but commercial firms are striving to reduce the cost by 90 percent.[67] If developed as a weapons system, they could dramatically change naval combat. Offensively, they can become self-deploying torpedoes or mines with transoceanic range. Defensively, they can be used to rapidly establish smart minefields in maritime choke points. They can be launched from a variety of surface and subsurface platforms or remain ashore in friendly territory until needed, and then launched from a port or even the beach.

Sea mines should be a particular concern to trading nations. Simple moored and floating mines have the distinction of being the only weapon that has denied the US Navy freedom of the seas since World War II. Mines have become progressively smarter, more discriminating, and more difficult to find. They have sensors that can use acoustic, magnetic, and other signals to attack a specific kind of ship.[68] And, of course, self-deploying mines can also be used against commerce. Launched from shore bases, these systems will allow any nation bordering, for example, the South China Sea and its critical straits to interdict trade. While they cannot stop trade, damaging a few ships will cause dramatic increases in maritime insurance rates as well as sharp decreases in the owners' willingness to risk their assets in the contested waters.

Air Warfare

For air power, the key problem will be protecting aircraft on the ground. An opponent does not have to fight modern fighters or bombers in the air. Today, China has developed the capability to destroy US headquarters, ships, runways, and most aircraft stationed in Japan with a combined ballistic and cruise missile attack.[69] Soon a much wider range of potential opponents will be able to dispatch hundreds or even thousands of small drones to hunt US aircraft at their home stations. Even if aircraft are protected by shelters, the radars, fuel systems, and ammunition dumps will be highly vulnerable. While the Flexrotor may seem too expensive to expend at $200,000 a copy, it costs less than operating a B-2 for two hours or an F-35A for just over four hours or firing one Javelin anti-armor missile.[70]

Manned aircraft will remain vulnerable due to basing issues even as cruise missiles and vertical launch drones become both more capable and cheaper. While the Tomahawk is an old system, it can serve as an example of what cruise missiles can do. A Tomahawk Land Attack Missile

(TLAM) cost $785,000 in FY 2013 dollars. If additive manufacturing can achieve the 40 percent savings Lockheed projects for satellites, a TLAM will cost about $470,000. These missiles carry a thousand-pound warhead for up to 1,500 miles (Block II).[71] While somewhat expensive, missiles such as these can provide long-range heavy strike—particularly if the warhead uses nano-explosives. Since they can be fired from a variety of land and sea launchers, they can be either dispersed or hidden in underground facilities (to include commercial parking garages and commercial ships) until minutes before launching. They will thus be immune to most preemptive strikes and much less expensive than ballistic missiles. The combination of cheap drones and much more capable cruise missiles provides relatively cheap, difficult to preempt, long-range, precision strike.

San Diego-based Kratos Defense and Security Systems Inc. has dramatically upped the ante in the field of drones with the XQ-222 drone. Its 1,500-mile combat radius is over two times the F-35's, and it can fly three thousand miles if sent on a one-way mission. It also has low-observable features, a five-hundred-pound payload, and no requirement for an airfield to launch or recover. It takes off with a rocket assist from a stand and lands using a parachute.[72] Kratos promoted the XQ-222 at the 2017 Paris Air Show, suggesting that the drones could fly in tandem with F-16 or F-35 fighters to dramatically increase the capabilities of existing aircraft.[73] Even at $2 million a copy, they are orders of magnitude cheaper than the F-35. And of course, their presence at the Paris Air Show indicates America will not be the only nation with this capability.

Modern drones and cruise missiles outrange modern fighter-bombers and thus can strike those aircraft at their easy-to-detect airfields—even expeditious ones. Thus they can push the vast majority of air bases out of range of their targets. While the United States has a capable tanker fleet, the refueling orbits will also be within the range of drones like the XQ-222. Today, only bombers can outrange drones. The fact drones can be launched from commercial, oceangoing vessels may also make bomber bases vulnerable.

Somebody has got to step up and make procurement efficiency the top priority. The F-117 [stealth fighter] development program worked well because there were only eight people in Congress who knew about it.   —Sam Nunn

Space Warfare

In space, the advent of micro- and cube satellites paired with commercial launch platforms will allow a middle power to develop an effective space program for surveillance, communications, navigation, and even attack of other space assets. Surveillance and navigation satellites are already within reach of most small or medium powers—or they could simply buy the services from a commercial provider such as Planet Labs as even the United States does.[74] Thus it will be hard to hide bases ashore or even moving forces at sea.

Cyberwarfare

While one would not normally think of drones as part of conflict in cyberwarfare, it is important to remember that all networks have nodes in the real world. Some are quite exposed. For instance, satellite downlinks and points where fiber-optic networks come ashore are known and vulnerable. Smart drones provide a way to attack these nodes from a distance.

Mass Returns

Since the 1980s, US forces have bet on precision to defeat mass.[75] This approach helped numerically smaller Allied forces defeat Iraq's much larger army (twice) as well as initially drive al-Qaeda and the Taliban out of Afghanistan using a very small ground force. However, technological convergence is pointing to the revival of mass (in terms of numbers) as a key combat multiplier. Additive manufacturing can make cheap drones fast enough that they can be used as rounds of ammunition.

How will US forces, which are dependent on a few exquisite platforms— particularly air and sea—deal with the small, smart, and cheap? Currently, the Defense Department is testing various directed energy weapons and electronic attacks to deal with the exponential increase in potential targets. However, like all weapons, directed energy will be subject to countermeasures. It is imperative that these systems be tested against a thinking, reacting red team that employs countermeasures such as autonomy, smoke, and shielding from electromagnetic energy. And one should keep in mind that land-based systems will have major advantages in power available and protection over any sea- or air-based systems. Thus, a directed energy system will also make domain denial easier.

The Return of Mobilization

After the fall of the Soviet Union, the United States abandoned the concept of mobilization. A primary driver was the fact that the US defense industry simply lacked the surge capability to rapidly equip a mobilized population. Mobilization in World War II was possible because industry could rapidly convert from civilian to military production. By 1990, the complexity of modern military weapons systems, plus the manufacturing plants and skills needed to produce them, made such a rapid conversion difficult if not impossible. Three-dimensional printing is inherently flexible since the product produced depends only on the materials the machine can use and the software that is loaded. Thus, as additive manufacturing assumes a greater role in industry, the possibility of industrial mobilization will reemerge. However, to succeed, mobilization must be planned to provide the necessary government-unique items such as warheads and target recognition software, quickly and in sufficient quantities.[76]

The Impact of the Fourth Industrial Revolution on Russia and China

Russia and China will be affected very differently by the fourth industrial revolution. Each will benefit, but neither as greatly as the United States.

Major investments, the free exchange of information, the free movement of people, and a relatively light regulatory environment seem to be necessary to maximize the benefits of this revolution.

Russia has none of these assets and is further restricted by severe domestic challenges. A fundamental problem for Russia is that its economy relies very heavily on oil exports, yet it cannot control the price. US shale oil is currently profitable at $50 a barrel on average, and some very productive plays have cost below $30 a barrel. America's ability to rapidly increase oil production in response to price increases means the price of oil is likely to remain in this range for the next decade or more. Despite OPEC's efforts to drive up oil prices, oil remained around $50 a barrel in mid-2017.

Compounding its problems, the Russian economy performs well below other European nations. In 2016, the Russian GDP of $1.3 trillion was below that of Italy despite Russia's population being 2.4 times the size of Italy's.[77] Even using purchasing power parity, Russia's 2016 GDP per capita ranked seventy-second in the world—just behind Greece.[78] Nor is Russia's economic future promising. To free itself from dependency on oil, Russia is trying to diversify by increasing manufacturing. Yet to do so, it will require major foreign direct investment. The lack of a robust legal system, lack of clearly defined property rights, and continuing sanctions have dampened the enthusiasm of foreign investors. While Russia had a positive FDI inflow of $10 billion in 2016, that did not compensate for the previous four years' outflow of $70 billion.[79] With Putin likely to remain in power—and with him, crony capitalism—-a surge in foreign investment is highly unlikely.

Thus Russia's economy remains stagnant and its population is declining sharply, with a projected drop of 10.5 percent by 2050.[80] Russian journalist Irina Grigoryeva has reported that the country is set to lose one million working-age people annually, resulting in an ongoing GDP growth rate hit of 0.4 to 0.5 percentage points.[81] Nor can Russia turn to a financial reserve to make up the difference. Because it planned its

2015 federal budget on oil at $100 a barrel, it has burned through its entire rainy day fund of over $90 billion.[82]

Even with all these problems, Russia will undoubtedly gain some benefits from the enormous increases in productivity created by the convergence of robotics, artificial intelligence, and 3D printing. However, Russia clearly lacks the political, economic, and social structures necessary to fully exploit the fourth industrial revolution.

Militarily, Putin's major effort to improve Russia's armed forces has clearly paid off. Quality and size have both increased over the last decade. However, that growth is coming to an end. Despite some highly publicized rollouts of new weapons prototypes and announcements of major expansions, the Kremlin actually cut its 2017 military budget by 25.5 percent. The cuts will remain at least through 2019.[83]

China is better positioned than Russia to benefit from the revolution but still faces major issues. Economically, China has been experiencing the normal middle-income slowdown of developing states. But it also faces a litany of major problems: a rapidly aging population, disastrous environmental conditions, massive public and private debt, a housing bubble, and the rising cost of labor. It must contend with all these problems even as it attempts to take the next step in development and shift from an export- to a consumer-based economy—all while trying to incorporate the changes driven by the fourth industrial revolution. Adapting is going to both help and hurt China. Economically, it will allow China to deal with the increasing cost of Chinese labor that has driven many manufacturers to cheaper Southeast Asian countries. In 2000, labor represented only 30 percent of China's manufacturing costs. By 2015, it was 64 percent.[84] Further, Chinese labor costs will continue to rise as China ages.

Chinese businessmen are fighting back by turning to advanced manufacturing—particularly robotics. The resultant massive improvements in productivity as well as dramatic reductions in personnel are allowing Chinese companies to remain competitive. To encourage this trend, China's government implemented its "Made in China 2025" program, which subsidizes ten key high-technology sectors. In 2016, Dongguan City created a plan to replace human workers with robots at 1,000 to 1,500 factories to strengthen its role as a manufacturing hub.[85] Foxconn reduced its labor force in a single factory in Kunshan from 110,000 to 50,000 by installing robots guided by artificial intelligence. The Changying Precision Technology Company replaced 90 percent of its labor force by installing robots and decreased its defect rate from 25 percent to 5 percent.[86] As many as six hundred other major Chinese companies have similar plans.[87]

While the large-scale deployment of robots is creating greater wealth for China, the massive loss of jobs is eliminating the path to prosperity for a large swath of its population. The Chinese Communist Party (CCP) is very sensitive about the potential for layoffs to cause instability and is aware of the increasing income inequality across China. Even as it is struggling to maintain its competitive edge with its private companies, it continues to heavily subsidize its state-owned enterprises (SOEs). The current effort to consolidate SOEs into a few large firms is not producing the hoped-for efficiencies. Instead, SOEs, which are granted preferential treatment by the government, are crowding entrepreneurs out of the credit market. A key issue going forward is whether the Communist Party continues to favor SOEs over entrepreneurs. That decision is likely to be driven by concerns more for stability than for efficiency.

The CCP is clearly deeply concerned about internal stability. Through 2013 it was actually spending more on internal security forces than on external defense.[88] In 2014, it stopped releasing the internal security budget, but many analysts believe it still spends more on internal than external defense. Given the fact the fourth industrial revolution can greatly increase the striking capabilities of insurgents, the CCP is likely to monitor closely and perhaps restrict the development of these new technologies as potential threats to stability.

China is already a world leader in some technology sectors and will benefit greatly from the fourth industrial revolution. However, the requirement to balance growth with control and stability will keep China from achieving all it could in these areas. It is essential to keep in mind that China is not a wealthy country. While it has the world's second largest economy, its huge population means that even measured with purchasing power parity, China ranks 102nd in GDP per capita—behind Costa Rica. And China will also be the first country to face the problem of growing old before it grows rich. By 2050, China's population will have decreased by 2.5 percent and have an age distribution similar to today's ultra-old Japan—but without Japan's wealth.[89]

> *On artificial intelligence, China has an ambitious goal. China wants to regain its place on the international stage, which they feel that they've been denied over the last 150 years.*
>
> —Karl Eikenberry

Great Power War

While Russia's nuclear arsenal remains an ever-present threat, it is one the United States has managed for decades. In contrast, Russia's economic, political, and social problems mean its long-term conventional threat to the West is declining. That said, Russia could seize select Eastern European states before sufficient NATO forces could arrive to stop them. The Baltic States, Finland, Sweden, and Poland recognize this fact, and all are taking steps to increase their defenses in the hope that will deter Russia. NATO is also looking at how to provide sufficient ground and air forces but understands their deployment will be too slow to stop a Russian invasion.

The creative use of swarms of autonomous drones to augment current forces could strongly and relatively cheaply reinforce NATO's deterrence. Deploying drones would increase Russia's uncertainty about its ability to execute an invasion while assuring that it will cost Russia much more to do so. If NATO assists frontline states in fielding large numbers of inexpensive, autonomous drones prepackaged in standard twenty-foot containers, the weapons can be stored in sites across the countries under the control of reserve forces. In addition, prepackaged IEDs could be built in similar and smaller containers using ammonium nitrate fertilizer. With a proper initiating charge, ammonium nitrate is a powerful explosive. A twenty-foot container can hold fifty thousand pounds of ammonium nitrate, which is safe as long as it is stored separately from the detonator. And the use of standard shipping containers massively increases the number of trucks available for moving weapons around the country as well as drastically complicating any attempt to preemptively destroy them.

Further, if the weapons are prepackaged and stored, the national forces can quickly deploy the weapons to delay the Russian approach. Even Russia's advanced forces could be faced with very large numbers of attack drones and dense networks of improvised explosive devices. Then NATO only has to fly in the relatively small crews necessary to augment those forces. If NATO also invests in long-range autonomous drones, Russia will face punishment from drones launching from essentially unlimited locations outside the range of its own tactical air force. And of course the Russian rocket launchers will also be subject to attack from the same long-range drones. The addition of GRAMM (guided rockets, artillery, missiles, and mortars) munitions, smart mines, and autonomous drones—and the creative use of improvised explosive devices— could make even small nations very hard targets.

Deterrence can be achieved through either denial or punishment. If an aggressor knows he cannot succeed or knows the punishment inflicted will exceed any gain, a rational aggressor will be deterred. The small, smart, and cheap approach can make deterrence affordable for NATO.

Against China, the small, smart, and cheap can also provide effective deterrence for the allies. To maximize the advantages, the allies must defend rather than attack. These weapons will dramatically reinforce a plan to hold the first island chain while denying China the use of the waters inside the first island chain or access to the ocean beyond.[90] The allies will not seek to win by striking into China but by choking her international trade and thus exhausting her ability to fight. This strategy, combined with new technology, could both deny China access to

the Pacific and Indian Oceans and punish it severely through an economic blockade. Thus it meets both requirements for an effective deterrent. Further, while the People's Liberation Army (PLA) envisions
winning short, "informationized local wars," the allies can establish a
defense that ensures it cannot win quickly.

A key strength in this approach will be land-based systems fighting
from the first island chain. However, to survive, today's allied air forces
and logistics facilities must be dispersed. Major bases are simply too
vulnerable to China's weapons.[91] US and Japanese forces have begun
limited exercises to demonstrate they can operate their air forces from
a wider range of bases. Given the level of threat today, it is essential that
US and allied forces regularly practice these dispersed operations.
Rather than operating from a very limited number of military airfields,
the air forces must practice operating from the numerous civilian airfields throughout Japan. Demonstrating this capability will have a
deterrent effect on China by greatly reducing the probability of a successful preemptive attack on forces in Japan. Naval forces can act as a
mobile reserve behind the chain to prevent Chinese efforts to break out
or overwhelm the defenses at a specific point.

However, dispersion only provides temporary respite. As long as
air power is tethered to bases inside the range of China's growing
arsenal of missiles and drones, it remains vulnerable to preemption.
It is essential that the United States and Japan cooperate in rapidly
designing and procuring long-range, advanced vertical takeoff and
landing strike drones. We already have promising designs such as the
XQ-222 and the Tern. The key is shifting investment from current
systems to accelerate the development and fielding of these systems.
When these systems are fielded, the basing possibilities become
almost unlimited and deterrence is increased because China cannot
contemplate a disarming conventional first strike. For the purchase
price of one Ford-class carrier and its proposed air wing, the United
States could buy ten thousand Kratos XQ-222 drones or thirty-three
thousand loitering TLAMs. It is pretty obvious which creates a greater
challenge for China.

Forces fighting on the defensive from the first island chain already have significant advantages against attacking Chinese air and sea forces. The most obvious is that Chinese forces will be fighting inside the allied air defense zone. The second is the fact that many of China's forces lack the range to reach the islands, and thus the allies will only have to fight a portion of the PLA. These advantages will grow as conflict shifts from few and exquisite platforms to the small, smart, and cheap. Using this strategy, the allies gain the advantage of thousands of independent, active hunters augmenting the relatively few but expensive weapons systems they currently own. Since it is easier to mass-produce drones in the range of one hundred to three hundred miles than those of five hundred to one thousand miles, the defense will have the advantage of numbers. And of course land forces inherently have much larger magazines and access to massive power infrastructures for directed energy weapons when they are developed.

As demonstrated by China's success with the construction of major naval and air facilities in the South China Sea, the US position is much less secure against Chinese "gray zone" incursions. The nations of Asia know that the South China Sea is simply more important to China than it is to the United States. Fortunately, all the nations involved understand the United States does have a major interest in freedom of navigation and will protect it, even in the South China Sea. Unfortunately, statements by President Donald Trump are contributing to the rising belief in Asia that the United States is no longer a reliable ally.

Conclusions

Unfortunately, the increasing reluctance of the US population to commit forces overseas, combined with the increasing cost of doing so in a conflict zone, does not bode well for America's ability to sustain our alliances. It will be necessary to reduce the cost to the United States and increase the visible contributions of our allies if we are going to maintain the exceptional strategic advantages our alliances provide.

The most important national security step we can take is to reinforce and reassure our allies in both Asia and Europe. As an aggressive and growing power, China can cause the most serious long-term disruptions in international security and the global economy. At the same time, North Korea is the most likely to cause a massive short-term disruption if it initiates a major war with South Korea. It would lose such a war, but only after massive devastation is inflicted on both sides.

Fortunately, in both Asia and Europe, US alliances are essentially defensive in nature, and therefore the technological advances favor the alliances. Unfortunately, alliances have tremendous inherent geographic and economic strengths but are unlikely to be sustained on the current trajectory. Converging problems with aging populations, infrastructure repair costs, medical costs, and surging military personnel and particularly procurement costs mean neither the United States nor its allies can continue along this path. However, if the United States takes advantage of emerging technologies, it can defend itself at much lower cost. Just as important, the lower costs mean the allies can assume more of the burden of their own defenses, which serves to reassure their populations as well as neutralize US complaints about free riding.

Working with our allies in Asia, the United States can provide deterrence against aggressive actions by either nation. Unfortunately, budget realities, not least the looming impact of rising federal debt interest payments, mean the United States must figure out how to assist allies in their defense in an affordable way. We can't continue on the path of buying fewer and fewer of ever more expensive weapons systems. We simply can't afford it.

Even as we continue to invest heavily in these legacy systems, China is refining numerous weapons systems specifically to attack their prime vulnerabilities: the need for major fixed bases and the limited range of most US aircraft. We are spending vast sums on systems for which China is already fielding increasingly effective counters. The first step the United States must take is to invest in further developing the new systems that will eliminate those vulnerabilities.

The United States has to maintain its current robust forward presence even as it begins to shift its war-fighting base from the few, vulnerable, and very expensive current platforms to the future force of smaller, cheaper, autonomous weapons that are not tied to major bases or sea platforms. Only by reducing costs can the United States maintain a viable strategic posture. A major benefit of shifting to small, smart, and many is that the United States can push its allies to follow suit. The deployment of a new generation of anti-ship cruise missiles, drones, and smart mines would greatly magnify China's challenges in either attacking another nation in the region or trying to maintain routes out of the South and East China Seas. Most important, these systems are simple and cheap enough that our allies could build, buy, and deploy their own in large numbers. A particular advantage is that large numbers of dispersed systems are not easily subject to interdiction. While China is building sufficient numbers of missiles and drones to gain confidence that it can destroy the few high-value US naval assets and the handful of Air Force and Marine air bases in Japan, it would not be able to gain that confidence about large numbers of containerized weapons stored in various locations across Japan.

Japan and South Korea both have sophisticated, capable defense industries. South Korea is already spending heavily on defense, so there it is a matter of shifting investment from legacy platforms to emerging ones. Currently, Japan, even with its support to US forces and its proposed defense spending increases, will not reach 1.5 percent of its GDP in the current ten-year plan. The United States must push Japan to at least achieve the NATO goal of 2 percent. It will be less difficult for Japan to increase its defense budget if the additional funds are spent on systems built in Japan by Japanese firms.

If Europe can overcome the rising nationalistic political environment, the fourth industrial revolution should bind its economy together more tightly as industry focuses on the regional market. The European Union has a GDP over twelve times the size of Russia's. Yet today, and for the immediate future, Russia has sufficient effective forces to be a serious threat to smaller nations in Eastern Europe. Thus for NATO the key question will remain one of unity. An essential element of keeping America tightly engaged in NATO is for those nations to demonstrate an increased ability to defend the alliance against Russia. NATO nations can do so by exploiting the cheaper but highly capable weapons evolving out of technological advances. Combined with thoughtful tactics and focused organizations, they can clearly deter and defeat Russian conventional forces. The United States needs to lead the way in the transition.

An important caution is in order. Allied strategists must recognize that for over two centuries wars between major, capable powers have been long—measured in years and even decades. The ability to replace combat losses is essential in a long war. With a peak planned production rate of only seventeen F-35s per month and several years required to build a new carrier, the United States cannot produce sufficient systems for combat replacement in a long war. It must start moving to a more survivable, sustainable, affordable, and replaceable force.

It is very likely that all nations—to include the United States—will have to contend with significant decreases in military spending. The potential for cheaper ways to accomplish the same mission means all nations should be very interested in exploiting the opportunities that are emerging. To be in position to exploit the new technologies and encourage our allies to do so, the United States must lead their development. Since the overwhelming majority of the investment is already coming from the commercial section, the focus for the Department of Defense must be to identify key niche technologies unlikely to draw commercial money—such as nano-explosives, smart mine fuses, and weapons that ensure the small, smart, and many have maximum combat power. The Russians have long owned small thermobaric warheads, and the Poles have successfully used them to arm man-portable, hand-launched drones. This kind of mash-up between old technology and new will greatly increase the effective combat power of each.

The biggest obstacles to the transition to a new generation of weapons systems and new methods of warfare will remain the political constituency that benefits greatly from the current systems and the inherent conservatism of senior military leaders. But it must be done.

To usher in this shift, the United States needs to develop a comprehensive plan for phasing in the replacements for its current extremely expensive and increasingly vulnerable weapons systems. Further, only if the United States leads the way in this transition will allied military leaders accept this new approach. Yet only by encouraging our Asian and European allies to rethink their procurement decisions can they afford to create real deterrent capabilities against Russia and China.

Like previous transitions (the carrier replacing the battleship, and precision munitions replacing dumb bombs), this one should follow the concept that the new technology first assists, then partners with, and finally replaces the old systems. We are well along on this path already in many areas. Today, drones have replaced manned aircraft for long-term surveillance in a low-threat environment. Cruise missiles are full partners—and often replacements—for manned aircraft. The Department of Defense has to make this a conscious procurement strategy rather than the current system of random evolution.

Today's transformations represent not merely a prolongation of the third industrial revolution but rather the dimly perceived arrival of a fourth and distinct one: velocity, scope, and systems impact. The speed of current breakthroughs has no historical precedent. In Klaus Schwab's words: "When compared with previous industrial revolutions, the fourth is evolving at an exponential rather than a linear pace. Moreover, it is disrupting almost every industry in every country. And the breadth and depth of these changes herald the transformation of entire systems of production, management, and governance."[92]

The United States' connectedness to the global economy will decrease, and thus the political will to bear the burden of providing international security—particularly in stability operations—will decrease. The high cost and perceived strategic failures in Vietnam, Somalia, Iraq, Afghanistan, and Syria have already significantly decreased the willingness of the American people to get involved in conflicts overseas. The continuing failure of allies to spend what many Americans perceive to be their fair share in their own defense remains an issue that angers many American voters.

The cumulative effect on the American public's willingness to stay engaged overseas has been remarkable. A 2016 Pew Research Center poll on the issue reported, "Nearly six-in-ten Americans (57 percent) want the U.S. 'to deal with its own problems and let other countries deal with their own problems as best they can.'" Just 37 percent said the United States should help other countries deal with their problems.[93]

At the same time, the fourth industrial revolution will provide smaller states and even nonstate actors with systems capable of inflicting significantly higher costs if the United States does choose to intervene. Further, the vulnerability of intermediate support bases will make the host nations much less likely to allow US forces to operate from those facilities except in time of very serious threat to the host nation.

The good news is that the United States is the nation best positioned to derive maximum economic gains from the fourth industrial revolution. US businesses are rapidly adapting to do so. The Institute for Supply Management announced on October 2, 2017, that its index of manufacturing activity was at its highest reading since May 2004.[94] And despite all the challenges noted above, the United States is in a good position to derive major national security benefits from the revolution. However, just as businesses cannot continue to operate based on old concepts, the government cannot provide for national security without making major changes to take advantage of the revolution.

3

GOVERNANCE AND SECURITY THROUGH STABILITY

Raymond Jeanloz and Christopher Stubbs

I think we've got to have a balance between optimism about what we can do with this technology but also realism about the dark side. —Sam Nunn

---

Just as advancing technologies are disrupting many sectors of the domestic economy, they are also transforming the international security arena. Diplomacy, deterrence, and direct military action, tools that have long been used to protect our national interests, are being challenged by rapidly evolving technologies that present new problems with no clear solutions. In sum, the power and pace of modern technologies call for developing new approaches to avoiding catastrophic conflict between nations.

In the past, the success of diplomatic attempts to manage such problems as the proliferation of nuclear weapons is in part attributable to the long timescales of years to decades typically involved with developing nuclear technology and weapon delivery systems. In the current era of ubiquitous, near-instant communication, however, today's rapid technological changes have greatly shortened the time for governmental consideration, decision, and action, even as these changes present complex and difficult new problems.

The governance challenge in this context is to identify and adapt to the interplay between technology and international relations. Equally relevant to the balance between technology and society are internal adjustments within each nation, from cultural and political to economic and technological, but responses should not be so onerous as to preclude technology from making its enormously positive contributions to individuals' and society's well-being.

We refer to this balance as "stability," a condition in which the pace of change—political, technological, and economic—is on a timescale that allows for governmental systems to adapt to those changes while maintaining sufficient equilibrium to allow steady progression toward prosperity. A disruptive technology is one that upsets this stability and induces changes on a timescale that is too rapid for a governance system to keep up. We use "crisis stability" to describe a condition in which sudden, localized events (e.g., regional conflict, natural disasters) do not drive the global system away from equilibrium, and propose here the notion of "technological stability" whereby the international system remains in equilibrium even in the face of high-impact, rapidly evolving—disruptive—technology.

The objective of this chapter is to explore the governance challenges of maintaining stability in the era of post–Cold War international relations and postmillennial disruptive technologies, with the intent of identifying specific objectives for stability. Stagnation, one form of complete stability, is neither beneficial nor desirable, in that change is both the cause and consequence of progress, whether in politics, culture, or technology. However, the goal is to reduce the chances of technologically induced catastrophe for society, including through inadvertent actions or unintended consequences. We illustrate the issues by starting with a brief discussion of nuclear weapons, a twentieth-century disruptive technology having uniquely destructive power.

Nuclear Weapons

Nuclear weapons, the most physically powerful military technology discovered to date, helped end one war and were supposed to change the nature of warfare forever.[1] That was not to be, however, as military conflict has continued relentlessly around the world. Instead, the recognition emerged that boundless, total war must be avoided because of the consequences of using nuclear weapons: the million-fold difference between nuclear and conventional explosives puts nuclear war in a special category of death and destruction.[2]

Warfare thus continues, but with a collective restraint to engage in combat as it was before the development of nuclear weapons. The effect of nuclear technology is to impose restraints, including through deterrence, to prevent attacks that threaten the existence of a state. Strategic alliances extend this deterrence to nations that do not themselves possess nuclear weapons.

During the Cold War, restraints developed in the context of strategic stability between the superpowers, the idea being to establish—among others—conditions for *crisis stability*, by removing incentives to initiate war, and *arms-race stability*, by removing incentives to increase the size or to enhance the capability of nuclear arsenals. The predominant form of Cold War deterrence was through threatened retaliation; for the superpowers, the primary function of nuclear weapons evolved to preventing their use. The overarching goal of strategic stability was thus to avoid nuclear war, the ultimate catastrophe enabled by the then newly discovered technology.

When you put your hand on the Bible, and swear to be president of the United States, that's the least of it. It's when you put your hand on the nuclear button—then you become God.
—Bishop William Swing

New Technologies

In the post–Cold War era the concept of strategic stability has become confusing, all the more so as a much greater number and diversity of actors are involved, from regional and global powers to terrorist and criminal cartels.[3] Is deterrence a reliable strategy under these circumstances? If so, how generally, and how is it to be implemented?

Moreover, the powerful new technologies now emerging give a sense that the pace of discovery is accelerating relentlessly. Technology is mostly viewed as positive—if not essential—for improving health and quality of life, often benefiting society by empowering individuals and increasing economic productivity. Yet many of these new technologies can also be used to inflict great harm. Do new technologies call for restraints and new codes of conduct between nations?

In particular, nonnuclear forms of deterrence show potential for addressing threats associated with modern technologies, notably deterrence through denial and deterrence through entanglement. The former amounts to avoiding or diminishing the effects of an attack, thereby reducing the incentive to attack. The latter—entanglement—is discussed below and may also be emerging as an important force in the nuclear domain.

Table 3.1 lists some of the important technologies that have come into prominence since the end of the twentieth century. The list is far from complete, if for no other reason than that new technologies as well as new applications of existing technologies are discovered every day. Also, it is somewhat artificial in that most technologies evolve in a continuous fashion, intertwined with other developments. For example, the global positioning system (GPS) plays a crucial role in supporting the internet and also in guiding autonomous vehicles, yet it is treated here as a twentieth-century technology and is left out of this listing. Similarly, nuclear technologies and genetically modified organisms are not included.

The majority of these technology developments have a dual-use nature, military and nonmilitary (or, depending on context, harmful and good). Commerce on the internet is accompanied by malware and

TABLE 3.1   Postmillennial Technologies

Information Technologies (IT)	Computers, smartphones
	Internet and Internet of Things (IoT)
	Artificial intelligence (AI)
	Social media
	Digital currency, blockchain technology
Biotechnologies	Genome editing (CRISPR/Cas9)
Space Technologies	Micro- and nanosatellites
	Robotic capabilities
Remotely Operated and Autonomous Systems	Remotely operated vehicles (ROVs)
	Autonomous air, underwater vehicles (AAVs, AUVs)
	Networked autonomous systems
	Robotic weapons

cyberattacks. Self-driving cars are emerging at the same time as robotic standoff weapons. Governments face the difficult challenge of promoting the economic benefits of these developments while protecting their citizens from the adverse impacts of the technology.

Information Technologies: Cybersecurity to Artificial Intelligence

The threats posed by information technologies, driven by enormous advances in computing speed and miniaturization as well as by the growth and pervasiveness of the internet, are recognized by the need to establish such government entities as the US Computer Emergency Readiness Team (US-CERT) and the Department of Defense's Cyber Command to provide defenses against cyberattack.[4] Moreover, the internet of things involves a massive linking of infrastructure that can greatly facilitate daily life but also brings considerable fragility to society and offers a multitude of entry points for damaging cyberattacks. The US government has identified "critical infrastructure" in an attempt to identify particular vulnerabilities to cyberattack.[5]

As is well known, the cyberrealm has already seen massive attacks and even acts of war through hacking, motivating immense efforts for defense of computer systems. In particular, a cyberattack puts critical infrastructure at risk, whether military command, control, communications, and intelligence (C3I) systems; the electrical grid; or banking and other financial structures. Some of these vulnerabilities are arguably of nation-altering proportion. Shutting down access to water, food, energy, and fuel for weeks across a large region could plausibly cause tens of thousands of premature deaths in a modern society, for example. Less dramatic but perhaps more insidious is the use of cyberattacks to undermine governance and stability within a society: the effects may cumulatively amount to an act of war, yet be too gradual to attract necessary public attention.

Potential consequences of cyberattacks include escalation to other domains and even initiation of nuclear war, if the strategic command, control, or communication systems of a nuclear-weapon state were hacked.

The international community is struggling to establish an international code of conduct in cyberspace.[6] The challenges of attribution and proportional response are complicated by instances in which the perpetrators are not part of any national structure. A technical challenge is to provide reliable and rapid attribution methods while preserving desirable aspects of internet connectivity. The balance between security and privacy in this domain is a topic of ongoing debate.

Artificial intelligence (AI), an emerging development in information technologies, shows promise to enhance and greatly speed up decision-making to the point that it is already used as part of many computer applications such as internet searches and identification and categorization of everything from words and phrases to images, video, and sound in electronic files.[7] While there have been breathtaking advances in recent years, AI's successes and—especially—failures are poorly understood. Many AI algorithms involve a large number of nonlinear mathematical transformations and can be prone to unintended or perhaps even unpredictable performance.[8]

Yet artificial intelligence is becoming encoded in numerous computer applications. It is conceivable—if not inevitable—that AI will be incorporated into systems that support critical military and civilian decision-making processes. The potential for accelerating and improving decision-making in complex environments may prove too tempting to resist, and AI could well be used in circumstances for which it is not suited or intended. This raises the possibility of key information being faulty or improperly analyzed, a recipe for disaster and unintended consequences should it ever happen in connection with nuclear or other high-consequence technologies.

A public debate is under way about whether and how governments should actively manage, regulate, and limit artificial intelligence. Some observers are calling for the imposition of restraints while others see this as unnecessarily hobbling an emerging new technology. A case has been made that the ethical, legal, and social implications should be part of the research agenda for artificial intelligence.[9]

Biotechnology

Biotechnology has exploded in capability and applications in recent years, holding great promise for improvements in medicine and agriculture. With its potential for industrial-scale genome editing, CRISPR/Cas9 exemplifies the modern revolution in this arena.[10] However, its power can in principle be turned to harmful use, notably in the development of unprecedented pathogens or other means to incapacitate humans or decimate agriculture.

To be sure, it has long been possible to create effective bioweapons, so it is fortunate that such weapons are not only banned by international agreement but also widely viewed as unacceptable and unjustifiable. Still, monitoring and verification are difficult, due to the small-scale and multiuse nature of biotechnology as well as the rapid pace of technology development.[11] To illustrate the point with an important contribution to medicine, live versions of the virus responsible for the 1918 influenza outbreak were reconstructed in 2005. The pandemic killed between fifty million and one hundred million people, far more than World War I, and its deadly character is now understood as a result of the research.[12]

Risk is no longer isolated—things nation by nation. It's everywhere. Someone gets sick in Nigeria, they're going to be in Chicago in twenty-four hours. It's fact.   —Lucy Shapiro

---

Space

The trend with space has been a sudden democratization due to commercial availability of rocket launches and the development of small satellites (micro- and nano-sats). Small nations, businesses, and even groups of private citizens are sending hardware into space, a domain that for decades had been accessible to only the largest of developed nations. As with information technologies, most of the expertise and hardware for space technology have shifted away from government control and exclusive access, now residing in the commercial sector.

Space and information technologies are thoroughly intertwined, such that a threat to one is a threat to the other.[13] For an increasing number of national governments, the loss of space infrastructure can mean the loss of intelligence, surveillance, and reconnaissance capability; of communications and control; and of medical, financial, and other infrastructure that depends on internet connectivity. Likewise, these space-based capabilities can be lost through cyberattacks.

Increasing robotic capabilities, such as satellites that can dock and service other satellites (perhaps using robotic grappling arms), can also pose a threat to national or commercial assets in orbit. Establishing shared expectations and rules of conduct, with a clear articulation of consequences for violating these international norms, would likely enhance stability in this domain.

As global society has become increasingly dependent on both space infrastructure and cyberinfrastructure, an attack on either domain can result in crisis. If nuclear-related space systems or cybersystems are attacked, a nuclear crisis ensues. More generally, all-out attack on space systems and cybersystems and the associated critical infrastructure could result in vast societal trauma, as large portions of the communications and transportation infrastructure are shut down, affecting everything from distribution of water, food, and electricity to provision of adequate medical care.

Remotely Operated and Autonomous Systems

Finally, remotely operated as well as autonomous air, sea, and land vehicles are being widely adopted for warfare and surveillance, in addition to numerous civilian applications.[14] This technology has emerged since the turn of the millennium, and we include it more as an indicator of the accelerating pace of development than because we understand its long-term implications. In particular, low-cost, highly capable autonomous systems can be networked to form powerful yet responsive swarms that could in principle be highly effective in either civilian or military applications. The important point for the present is that there are likely to be major consequences of these developments, but they are not well understood at present.

We used to belittle the Chinese and think we always had ten years on them technologically. Now, as you point out, they're right behind us, and oh, by the way, they tell us, "We no longer copy you anymore."   —James O. Ellis, Jr.

Strategic Stability

Here, we retain the Cold War goal of strategic stability, to avoid nuclear war, and broaden it in two ways so as to address the disruptive consequences of newer technology.[15]

First, we acknowledge that certain attacks through other, nonnuclear technologies could trigger a nuclear response. Historically, US policy has left open the possibility of nuclear retaliation to the use of any weapons of mass destruction (WMDs), including chemical or biological weapons. There is also the potential for nuclear response if nuclear forces or nuclear command-and-control systems are attacked by nonnuclear means, whether with conventional arms or by way of new technologies (e.g., cyberattack). The conclusion for maintaining nuclear strategic stability is that it is beneficial, if not essential, (1) to separate nuclear from nonnuclear military infrastructures; and (2) to avoid attack on any nuclear command-and-control systems.

Second, the impact of the new technologies is not limited to their physical power, but includes the overall consequences of their use in aggression. Modern society has developed efficient and effective infrastructures of material objects, human activities, and relationships between all of these. Yet this interconnected web is fragile and therefore presents key vulnerabilities. Thus, we have posited, in examples given above, modes of attack that would bring a country to its knees through massive shutdown of critical and technically fragile infrastructure. Some of these attacks could plausibly cause enormous numbers of deaths, perhaps not as quickly as nuclear attack but nonetheless effective at crushing a society, more likely over days and weeks rather than in minutes to hours.

This amounts to a threat by an external agent to the continued existence of at least a nation's social and political system, albeit not likely to every human life in its population.[16] Consequently, more than one leading scholar of modern technologies and defense identifies cyberattack as the greatest threat to the United States and other nations over the next decade or so, based on expected value of damage.[17] In this regard, the harmful use of biotechnology is another threat with catastrophic potential.

We acknowledge at the outset that far more analysis is needed on the potentially disastrous consequences of technology. Nuclear weapons stand unique in terms of their capacity for causing physical destruction, representing from a technical perspective the preeminent weapon of mass destruction. Still, realistic assessments of societal impact are needed for all of the modern and emerging technologies, with the primary focus being to diminish the possibility of their catastrophic misuse.[18] In some cases, such as the cyberdomain, the possibility of misuse may be reduced by mitigation of the effects of an attack: resiliency can provide a certain level of deterrence by denial.

Timescales for Instability

With these considerations in mind, we use timescales to distinguish two elements of strategic stability: *crisis stability* and *long-term stability* (table 3.2). The goal of crisis stability is to avoid an existing crisis between two nations from escalating to a catastrophic level. In principle, this form of stability should include a reliable means of de-escalation from the significant crisis. In contrast, a crucial aspect of long-term stability is to maintain governance that reduces the likelihood of potentially catastrophic crises arising in the first place.[19] In sum, long-term stability should provide a solid foundation for efforts to achieve crisis stability.

Short timescales, high stress, inadequate situational awareness, broken channels of communication, and the potential for misunderstanding,

TABLE 3.2 Elements of Strategic Stability

Crisis Stability	Avoid incentive to initiate major conflict from crisis
	Establish and maintain decision-making integrity
	Maintain situational awareness for both sides
	Establish and exercise means of de-escalation
	Slow down pace of response and decision-making
Long-Term Stability	Understand new technologies, advance technological stability
	Establish norms and promote control of technologies
	Enhance resilience
	Avoid incentives to develop new threats
	Promote entanglement

miscalculation, or miscommunication are key challenges for crisis stability. Both internal and external communications are at risk; the first to maintain a reliable chain of command and responsibility within each nation, and the second to provide for clarification and negotiation between the opposing states.

In comparison with crisis stability, long-term stability requires political focus, commitment, and support over extended periods of time, the goal being to establish effective procedures for avoiding major crises. Enhancing confidence between potentially adversarial states is among the features of long-term stability. Ironically, success in eliminating significant crises can lead to a loss of attention by one side or the other, to the detriment of both. Clearly, internal politics plays a role in each nation's level of long-term commitment to actions that enhance strategic stability. The governance structure and protocols that achieve and sustain stability need to be established through joint discussions among all interested parties and cannot be prescribed by a single nation.[20]

Crisis Strategic Stability

Under crisis stability, deterrence is traditionally considered an important means of removing the incentive for a nation to strike first and initiate major conflict. Many regard this to be the primary if not exclusive role of nuclear weapons, for those states that have them.[21] Of course, having a strong conventional military capability can also play an important stabilizing role, in that conventional is far less risky than nuclear response and can, under the right circumstances, provide an effective deterrent.

For the new technologies discussed here, deterrence is compromised by the potential difficulty of rapidly and reliably identifying the perpetrator(s) of an attack. In many instances, this difficulty stems from the technologies being widely available and multiuse, as well as powerful.

Cyber

Rapid and reliable attribution can be difficult for cyberattacks, the result being that deterrence by threat of retaliation may not be credible. One deterrence option under such circumstances is through denial, removing the incentive to attack by reducing either or both (1) the chance that the attack can succeed and (2) the damage that it can inflict.[22] This requires establishing effective security, increasing resiliency and redundancy of the systems to be protected, and making other demanding efforts.

How much is enough, however? Much more study is needed to answer this question, but we can imagine different levels of fortitude that involve distinct levels of collaboration. Protection from cyberattack by nonstate actors such as criminals and terrorists can be viewed as a common good addressed by international cooperation, for example. Successfully blocking such attacks may therefore be a reasonable objective to be pursued through international bodies, such as the United Nations or Interpol.

On the other hand, the best that one might do in response to attack by a major state is to reduce or slow down loss of capability, for instance in cyber or space domains, a form of deterrence by denial. Alliances may help in quickly replacing communications or satellite infrastructure, but one would expect less opportunity for avoiding complete loss of capability, in comparison with attacks by individuals or small groups.

Deterrence in the cyberdomain is thus expensive and incomplete, but one aspect (denial) has the positive attribute of improving system reliability overall. For instance, resiliency and redundancy make it possible for infrastructure to perform (perhaps at reduced levels) despite natural catastrophes or other malfunctions, whether due to aging components or human error. Of course, the *possibility* that there will be rapid attribution and retaliation can also play a role in deterring external cyberattacks.

On cyber, we should not believe that even if we had an excellent international agreement, that it would deal with the whole problem, or even a majority of the problem. The biggest problems here are not nation-states but groups outside the law. —William J. Perry

---

Space

Commercial and military space-based communications and surveillance capabilities have historically become concentrated in small numbers of expensive, highly capable satellites. The effective functioning of our civil societies and military establishments depends critically on these legacy space assets, which are increasingly vulnerable to new technologies (e.g., cyber) and more conventional threats (e.g., kinetic).

It may be possible to build more resilient space-based capabilities for the most important military and civilian purposes, drawing in part on small satellite technologies. Small satellites can be constructed and deployed far more quickly than large satellites, and reduced capabilities may be at least in part mitigated by larger numbers being rapidly put into orbit. Careful analysis of the costs and trade-offs involved is warranted, as increasing the resilience of space-based systems could make a substantial contribution to crisis stability.

The capabilities of commercial satellites have improved dramatically in recent years, so there is the possibility that these can offer some level of backup for national systems in times of crisis. It is interesting, in this regard, that nongovernmental experts have used commercial satellite imagery to make significant contributions in areas of surveillance traditionally dominated by major nation-states, for example in nuclear nonproliferation and treaty verification.[23]

Command, Control, and Situational Awareness

Establishing and maintaining decision-making integrity generalizes the function of a national command authority for nuclear weapons, so that it applies to all dangerous technologies. Instituting a robust chain of command and avoiding predelegation of the decision to use nuclear weapons to anyone other than the highest authority is the standard for major nuclear-weapon states. Extending that concept to other modern technologies, it would be a matter of stated policy that the highest authority is required for any major attack, whether nuclear, cyber, or otherwise. Also, reliable and independent channels of communication should be available to the highest command authorities. It is for both nations' sake that these communication channels not be attacked so that they can function properly in times of crisis.

The difficulty comes with dual-use technology, such as space and cyber, as well as with technology that can be used at many levels of severity (e.g., from low-consequence hacking to crippling cyberattack). Out-of-control rogue elements can be highly destabilizing if they successfully launch a major attack, for instance through cyberintrusion. Therefore, it is in the interest of each nation that such actions be treated either (1) as attributable to a national authority; or (2) as a terrorist (or criminal) threat to all.

Similarly, safety and security are essential to ensuring that neither accident nor third-party actions cause a nation to initiate a catastrophic attack. Nuclear-weapon safety, for instance, is intended to prevent accidental detonation of a nuclear weapon due to a nearby explosion (such as for a nuclear weapon located in a region of conventional conflict), the point being to avoid a nonnuclear event inadvertently becoming a trigger for nuclear conflict. Safety and control mechanisms over biological, space, and other technologies can play an analogous role.

It benefits each party in a crisis to ensure situational awareness for *both* nations. This implies avoiding destruction of crucial intelligence, surveillance, and reconnaissance (ISR) systems and their associated communications infrastructure.

The alternative is that either or both nations may be forced to make decisions based on incomplete or unreliable information, with a potential—if not likelihood—of making worst-case assumptions. This is a recipe for escalation and for a crisis spinning out of control. In the case of nuclear weapons, the objective for stability is to avoid either nation being in a use-or-lose position due to unreliable information and to reinforce the proscription against attacking nuclear systems even by nonnuclear means. In the context of space systems under attack, significant degradation of a nation's communications, command, or ISR capability may trigger major retaliation by that nation, before all functionality is lost.

I think if the public understood this system, and how fragile it is, and how things are susceptible to possible human error—I think we'd have demands for changes on this.   —Sam Nunn

---

Communication, De-escalation, and Pacing

Actions providing general conditions for enhanced crisis stability are important, but actually dealing with a crisis requires communication between (at least) the two states involved. This underscores the need for a hotline between the nations' leaders and a well-understood chain of command on both sides. It is not just a matter of installing emergency channels of communication, but also of exercising them in order to provide assurance that they will function in a time of crisis. Procedures for recognizing and containing crises as well as for de-escalation are worth working out ahead of time, especially if a third party is involved (e.g., two nations responding to a major cyberattack from one into the other, but actually perpetrated by a third nation).

Finally, we note that the essence of crisis stability is to lengthen the time available for observation, decision-making, and negotiation; that is, to slow down the pace of events to the degree possible. As historically developed in the nuclear realm, hotlines and well-established procedures for crisis response can play an important role for pacing as well as informing decision-making and negotiation. This is a realm, however, in which we imagine future integration of AI could be problematic because of its potential to speed up timescales for decisions and because of the added possibility of injecting faulty information or analysis into the crisis-management process.

Many if not all of the restraints proposed above have been considered in the nuclear realm, but less so (or not at all) in the context of the newer technologies.[24] However, to the degree that these technologies are powerful, multiuse, difficult to attribute, and rapidly becoming ubiquitous, it is essential to examine the destabilizing consequences of their use, especially in offensive actions, lest those technologies all too quickly come back to haunt those who have invented or first deployed them.

I draw a parallel between cyberattacks and the issue of weapons in space—that is, in orbit. In both cases, we've understood these threats, and people have been warning about them for a long time. But every time the issue comes up in Washington, it is dismissed—either legally binding obligations or even the code of conduct—on two grounds. One is that we couldn't verify any sort of limitations. But the other argument, the one that carries the day, is that we, the US government, want a free hand to do this ourselves, and we can win any competition in this area—which I think is mistaken.   —William J. Perry

Long-Term Strategic Stability

Long-term stability is a generalization of arms control and counterproliferation in the nuclear realm. Its goals are building confidence; reducing the likelihood of crises that could precipitate conflict; and providing a basis for crisis stability between nations. History shows that efforts must be sustained over long periods of time in order to achieve success in arms control. We expect the same in applying the analogy to newly emerging technologies. In some sense, it is the very process of establishing norms, seeking common ground, and otherwise establishing entanglements (see below) that builds confidence between adversarial nations.

Achieving this goal is not easy. It requires as thorough an understanding of emerging technologies as possible. Understanding the consequences of their use is as important as determining what is needed for their development. A responsible government will want to evaluate the potential threats as well as the benefits from any powerful new technology, so it should ensure that the requisite technical studies are carried out by government, academic, and commercial sectors.

A case in point is Einstein's August 2, 1939, letter calling President Roosevelt's attention to the newly discovered process of nuclear fission.[25] This is not to imply that powerful new technologies ought to be, or necessarily will be, developed for military applications. Quite the contrary—technical study provides the basis for understanding both the need for and means of controlling new technologies, so as to maximize the benefits and minimize (if not remove) the threats posed by these technologies. This balance between benefits and threats and between disruptive impact and society's response lies at the heart of technological stability, to which we return below.

A key question regards potential blowback from military applications of new technologies. There is the utilitarian question of how quickly and effectively others can adopt and even excel with the technology, such that its development and use in a military context is ultimately disadvantageous. Given both the power and ease of acquisition of the newest technologies, this pragmatic issue becomes more important than ever.

In addition, there are ethical concerns about any powerful technology: just because it is available does not mean it should be used. More specifically, how a new technology is to be used, and the underlying norms for its use, need to be thought through for any military application, lest the consequences of use ultimately outweigh the benefits. The ethical and moral implications of new technologies are at least as important as the technical and pragmatic considerations that we have emphasized in the present discussion, especially as military action is as much a projection of a nation's values as a projection of its power.[26]

Objectives

Four objectives stand out for long-term stability in the light of new technologies: (1) establishing a set of norms for use of the technologies; (2) building resilience into civilian and military infrastructures; (3) reducing incentives for the development of significantly more powerful threats in any given domain (i.e., avoiding vertical proliferation); and (4) providing incentives for controlling the harmful or threatening aspects of the technologies. The last is a modification of the established notion of horizontal proliferation, to account for the fact that many of the powerful new technologies are widely accessible and have multiple uses, including numerous beneficial applications. That is, the assumption has to be that the technology is, or will soon be, widely available to other nations. Biotechnology is clearly in this realm now, and it is plausible that nuclear technologies will become systematically more accessible over time.

One means of developing norms and building confidence that has proved effective is for nations to maintain regular consultations between governmental counterparts, including between military counterparts. These dialogues can be expanded to incorporate tabletop exercises, for example to practice crisis control, and can be supported by unofficial (Track II) meetings in case the two countries are unable to discuss certain topics for political or other reasons.

Agreements on how to handle dangerous military activities, such as incidents at sea and in the air, are also stabilizing, as are advance notifications of military exercises and missile-testing launches. These serve as confidence-building measures and, more explicitly, reduce the chances of an event inadvertently becoming a crisis.

Public statements of policy are also important, whether to explain decision-making processes and topics of concern or sensitivity or to clarify a nation's position on uses of technologies. Adopting a policy of defensive last resort or of no first use is one example from the realm of nuclear weapons. Analogous clarifications need to be made about a nation's chain of command and leadership responsibility for the most powerful of the new technologies. Use policies need to be spelled out, whether for lethal attacks with drones or for the panoply of defensive as well as offensive cyberoperations. Notably, some cyberdefense operations are indistinguishable from, or can be easily misperceived as, acts of offense.

More generally, public rejection of terrorism as having any legitimacy is an important component of long-term stability, because blurring the lines between national and subnational activities is especially dangerous in a world of powerful technologies that can be used by individuals or small groups rather than requiring the commitment of an entire nation.

We point to the unintended consequences of the 9/11 attacks, which were apparently far more successful in destroying their targets in New York and Washington, DC, than the perpetrators expected. Hundreds of thousands have ultimately been killed in Iraq, Afghanistan, and elsewhere because of that single day's attacks. The important conclusion is that many of the new technologies emerging at the start of the present millennium are more powerful than those of past centuries. The unintended consequences, and perhaps even intended consequences, of using these new technologies are poorly understood.

When you listen to the discussions now of major political figures, it's pretty clear that they are talking on a very thin and superficial level about how to handle problems. —Charles Hill

Entanglement

Harvard scholar Joseph Nye emphasizes entanglement as offering another form of deterrence based on countries sharing interests in the modern world.[27] Globalization and the large-scale migration of populations from one nation to another—for education, job opportunities, or other reasons—provide one level of entanglement between nations.[28] Economic relations represent another form of entanglement, whether one state views the other as an emerging market for sales, a source of cheap but capable labor (e.g., for manufacturing), or an opportunity for broader investments—from real-estate development to capitalizing the other nation's debt. All of these investments and relationships offer reasons for maintaining stability and avoiding war.

Entanglement arises from common interests, for instance when (potentially adversarial) governments realize that it is mutually beneficial to prevent money-laundering. Sometimes, there are even shared norms among a wide diversity of cultures, including nations that otherwise have conflicting values: intolerance for child pornography and rejection of chemical and biological weapons are to a large degree in this category. To the surprise of many in the United States, concepts of free speech are more nuanced, with disagreement about appropriate bounds for speech (e.g., "hate speech," political activism, calls to violence) being evident even among allies sharing significant cultural history.

The relationship between the United States and China is an example of entanglement, whereby both countries are economically interdependent to a degree that stabilizes otherwise conflicting—sometimes antagonistic—military, political, and cultural agendas. It would be profoundly painful to individuals, businesses, and the governments of *both* countries were either to initiate war with the other.

Nye acknowledges the false hopes of a century ago, when world trade was thought to make large-scale war impossible, only to be proved wrong by the outbreak of World War I. However, he makes a case for present circumstances being different, with ubiquitous and near-instantaneous communication bolstering an international web of deeply connected economic, cultural, and even political relationships. In any event, while there can be questions about the magnitude of the effect, entanglement with a potential adversary complicates any decision to initiate the use of force and thereby contributes to deterrence.

It is not just that globalization encourages entanglement but that technology facilitates—if not depends upon—the interconnections that lead to entanglement. Global, ubiquitous, near-instantaneous communication both leads to and is a result of entanglement around the world.

It can also be thought of as a confidence-building measure. A case in point is the use of space, which becomes more secure against rogue actions the more that nations become dependent on—and therefore invest in—the services provided from space. Notably, states that depend on space are averse to the generation of orbital debris from kinetic attack against space infrastructure, even if not their own hardware; aside from loss of services, the debris can threaten current and future satellite orbits. Mutual interest thus becomes the basis for restraint, amounting to both establishment of a norm and deterrence against violating that norm.

To be effective, entanglement should (1) involve as broad a spectrum of the two nations' economic and political interests as possible; (2) engage these sectors as deeply as possible; and (3) respond to potential instability and crisis in as timely a manner as possible. This calls for a partnership between the public and private sectors associated with the new technologies now emerging. More generally, entanglement along with resilience (i.e., deterrence by denial) may in combination provide the best means of avoiding technologically induced catastrophe between nations.

Technological Stability

Our discussion has emphasized the threatening and destabilizing consequences of technology resulting from the potential imbalance between the magnitude and pace of technological change on one hand and the capacity of governance, culture, and politics to accommodate those changes on the other. As the evolution of technology is notoriously hard to predict, governance systems are often playing catch-up after major technological changes are well under way. Technological disruption can be both beneficial and harmful, however, and it is the positive contributions that are—in most cases—the motivation for development of technologies.

It is therefore essential to develop and support technology's benefits while evaluating and learning to control the threatening consequences of its existence and use. This is all the more true because one nation deciding to limit use of a technology does not mean that others will follow suit, especially if there are immediate benefits for adopting that technology.

TABLE 3.3   Technological Stability

Technological Stability	Expand access to positive aspects of emerging technologies
	Protect citizens from adverse impacts of emerging technologies
	Adapt to and exploit rapid technology advances so as to retain stability
	Deter misuse of emerging technologies: denial, entanglement (retribution)

The internet and biotechnology illustrate that what is considered to be the most appropriate and beneficial control of a technology can be a matter of considerable disagreement, even among like-minded nations. The multidimensional balance between the benefits and threats of technology—and between its disruptive character and societal accommodation—is what we refer to as technological stability (table 3.3).

General goals are clear enough, but the most effective means of reaching these goals needs far more analysis and development. Details matter, and it may turn out that each technology will require specialized or even unique approaches to achieving the analogue of arms control, deterrence, and nonproliferation that has served the nuclear realm.

To be sure, not all technologies are threatening, and there is much opportunity for enhancing stability by cooperatively developing and applying such benign technologies. More efficient distribution and use of energy could reduce reliance on vulnerable grids; small satellites could reduce reliance on vulnerable space assets. These are but two examples of the means by which new technologies could be exploited for enhancing stability.

Conclusion

In an international economy in which growth is essential to prosperity, a static system is neither stable nor sustainable. Change is inevitable and essential. Improvements in technology have played a central role in increasing productivity and improving the quality of life. The governance challenge is to capitalize on the positive aspects of technological progress while guarding against adverse effects, intentional or otherwise. Adverse effects can include not only misuse but also suppression of technology.

There are many examples of information technology being abused, both in the criminal domain and cyberwarfare. Fortunately, we have yet to see an example of widespread abuse of modern biotechnology.

Establishing shared expectations and general principles that can be applied to each of the emerging technologies, in order to deter their misuse or abuse, could lay a foundation for a stable system. Establishing and clearly declaring national policies that conform to these principles could facilitate steps toward a new deterrence regime that would foster stability while allowing for evolving technologies to continue to be a force for positive change.

Specifically, as nations become increasingly reliant on sophisticated systems that are fragile—satellite constellations, electrical grids, computerized financial systems, and more—we need to establish an international security regime that effectively deters attacks on these systems. This regime can include increasing resilience and reducing vulnerabilities, while taking appropriate steps to sustain these systems' societal benefits. Mutual interdependence on these technologies and their associated commons leads to an entanglement that in itself can be stabilizing. In short, some of the challenges of technology—its power, pervasiveness, rapid evolution, ubiquity, and ease of access—can potentially offer a path to new, more resilient infrastructure and to new modes of international cooperation that enhance stability.

What is to be done? We're so far from having answers to that question. Many people in Washington don't even have the question yet.   —Niall Ferguson

---

4

GOVERNANCE IN DEFENSE OF THE GLOBAL OPERATING SYSTEM

James O. Ellis, Jr.

Technology is a useful servant but a dangerous master.
—Christian Lous Lange (Nobel Peace Prize, 1921)

Nearly one hundred years ago, Christian Lange was awarded the Nobel Peace Prize for his lifelong commitment to and advancement of the theory and practice of internationalism, the promotion of greater political or economic cooperation among nations.[1] In his Nobel address, Lange presciently described the essence of the challenges still confronting our world a century later: "Today we stand on a bridge leading from the territorial state to the world community. Politically, we are still governed by the concept of the territorial state; economically and technically, we live under the auspices of worldwide communications and worldwide markets."[2]

Author's note: I would like to acknowledge the significant research contribution to this chapter by Alexander Stephenson while he served as a research intern at the Hoover Institution in the summer of 2017. This chapter builds on the author's 2016 Hoover Institution conceptual essay, "Readiness Writ Large," and draws on a 2016 co-chaired National Academies report and subsequent congressional testimony on US national security space defense and protection.

Lange devoted his life to the pursuit of an unrealized and, increasingly, unrealizable dream—the unity of mankind.[3] His work began before the First World War with the internationally collaborative Interparliamentary Union. But, energized by the horrific impact of the conflict, in its aftermath he aligned himself with and represented Norway in the nascent, and ultimately unsuccessful, League of Nations, seeing in it the poet Tennyson's confidence, "In the Parliament of man, the Federation of the world."[4]

Lange's concepts have been updated in recent decades with descriptions of new world orders that have proved in reality to be anything but orderly. It may be more useful for our discussion to advance the concept of the "global operating system," drawing intentionally on computer terminology. This serves to underline the linkage that technology enables in our hyperconnected world and the theme of this book.

An "operating system" is a computer's underlying software framework. It manages the hardware components and enables all other programs and applications. Its responsibilities scale with the complexity of the system: policing activity such that simultaneous users or programs *do not interfere* with one another, ensuring security whereby unauthorized users *do not access the system,* and *establishing formats and standards* that must be met by all applications and users. There are clearly parallels with the myriad banking, market, communication, information, and national security networks that are the hallmark of our globalized world. In a recent *Foreign Affairs* essay, Tim Kaine adds that when travel, information sharing, technology, immigration, and commerce are added, nations are drawn and held together far more closely than ever before. And the post–World War II system of international norms, rules, and institutions—a system the United States played a major role in building—draws countries closer together still.[5]

In a sense, the term "global commons," typically used to describe international resources—including shared natural resources like oceans, the atmosphere, and space—has been expanded to include cyberspace. But this has occurred without the societal norms, governance standards, and security expectations and capabilities resident, for the most part, in the other domains. For my purposes, let us define the global operating system as including not just the neurons, blood vessels, and connective tissue of our global body politic, but also all the data, information, knowledge, and actions that are transmitted on or enabled by them.

This operating system has evolved to include the technologies, customary behaviors, conventions, and, eventually, treaties governing diplomatic, military, and commercial activity. It also includes operational concepts, strategies, nation-states, and attendant diplomacy that enable those norms. Because of its slow, evolutionary character, the system has up to now been able to gradually incorporate technological advances, accommodate stresses, and, to some degree, resolve conflicts in a deliberate manner over time.

In our era, however, the speed of recent advances in global connectivity and technologically enabled capabilities has significantly outpaced the creation of guiding national strategies and policies. The technological advances in—and increased national security reliance on—the global operating system have created a common global "critical infrastructure" that has not been matched by coherent supporting protection and loss-mitigation strategies, clearly articulated policies, and robust defensive capabilities. These gaps have created newfound concern domestically, confusion on the part of allies, and misalignment and misperceptions on the part of potential adversaries. We must now urgently fill policy gaps, find mitigation strategies, and establish new defensive capabilities.

That the global operating system is under siege is unarguable. Even if the perpetrators remain hidden, the evidence is clear, from large-scale cyberattacks to the massive online theft of billions of dollars in intellectual property; from the rampant insertion of trust-sapping "fake news" to the attempted interference in democratic processes around the globe; from shadowy probes of critical infrastructure control systems to highly visible attacks on commercial entities; from unattributable espionage attempts to well-coordinated hybrid warfare inciting cross-border social unrest; from confrontational encounters in the global commons to attempts to fracture decades-old alliances and partnerships. At every level and in every domain, the number and pace of attacks are growing.

Background

Christian Lange, cited above, was part of a centuries-long parade of, first, European and then global advocates for a broader and more effective world order. By reviewing some of those historic efforts, we can gain some insights into the challenges, changes, and lessons of the past that can inform our way forward. To paraphrase Pavlov: "If you want a new idea, read an old book."

Many mark the beginnings of this effort with the Treaty of Westphalia ending the Thirty Years War, a conflict that Henry Kissinger describes as "a conflagration in which political and religious disputes commingled, combatants resorted to 'total war' against population centers, and nearly a quarter of the population of Central Europe died from combat, disease, or starvation."[6] The Westphalian treaty of 1648 enshrined the concepts of a balance of power among signatory states, national sovereignty, and noninterference in another country's internal affairs. As Kissinger further emphasizes, the Treaty of Westphalia was not conceived as a globally applicable system, not because it did not have that potential but, pragmatically and importantly, *"because the then-prevailing technology did not encourage or even permit the operation of a single global system* [emphasis added]."[7]

Despite the significance of the Treaty of Westphalia, a very different group of statesmen assembled at the Congress of Vienna in 1814 to deal with the wreckage of the order created so optimistically a century and a half earlier. Richard Haass succinctly summarized both the imperative and the outcome: "The leaders of the day were so traumatized by what had just taken place that they operationalized the concepts of the Westphalian model, resulting in the Concert of Europe. The concert, as the word suggests, was an orchestration of how international relations in Europe would be conducted" and required the restructuring of a new balance of power from the wreckage of the old, while accounting for both the rise of nationalism and the impacts of the fall of France and the rise of Russia.[8]

Despite the successes and longevity of the Congress of Vienna, its structure and relationships failed in the first half of the twentieth century to prevent two horrific world wars. The causes of World War I and World War II were many and very different. Much scholarly research and countless books have detailed the political, economic, and societal costs and described linkages between the two. Fulsome discussions are beyond the scope of this chapter, but there are certainly lessons to be learned that can and should inform our thinking, especially as we work to strengthen and defend the global operating system and if we are to avoid future large-scale conflict.

Haass sees two main lessons from World War I that may be applicable to our review. First, he notes that world orders, or global operating systems, for that matter, are not automatic or self-sustaining, even when they are patently in the interests of all who benefit from them. The war benefited no one and cost the protagonists far more than they gained. There are real limits to enlightened self-interest and a real balance of power; despite their existence, they were not enough to keep the peace. For our purposes, this argues that creation and sustainment of a global operating system require constant attention and adjustment as situations and circumstances inevitably change. In aviation parlance, there is no autopilot that automatically corrects for what, initially, are small perturbations that, left uncorrected, can spiral out of control. This fact will bear on our subsequent discussion of an appropriate US role in shaping and maintaining the global operating system.

A second lesson is the limits of economic interdependence. Before World War I, trade was flourishing, but the widespread mutual economic benefits did not prevent the conflict. In his 2015 work *Economic Interdependence and War*, Dale C. Copeland finds that the issue is much more complex. The so-called liberal belief that trade and investment inevitably reduce the likelihood of conflict is not always true. He cites as but one example the Japanese fear of loss of resource market access as a proximate driver of Japanese expansionism and aggression.[9] In a sense, Japan's reliance on international trade encouraged rather than discouraged conflict. Here, too, is a lesson that will need to be considered as we attempt, for example, to refine or redefine our relationship with China.

There are lessons, as well, from World War II, one of which bears particularly on today's nationalist, if not isolationist, trends. Haass opines that European and American actions (or lack of them) were responsible for World War II. Here he specifically calls out the unrealistic and unrealized interwar hopes placed in the League of Nations, the failure of political consensus that prevented American participation in this version of "the new world order," the retreat by America into isolationism (weakened and distracted by the Great Depression), and, finally, the policies of appeasement and disarmament that dramatically shifted the balance of power.[10] Here, too, there may be powerful lessons about leadership in actively helping shape outcomes rather than passively accepting what the future brings. The ability to craft a compelling, inclusive narrative will be essential in defining a way forward. As George Shultz often notes, "You have to be onboard at the takeoff if you want to be there at the landing."

In contrast, the post–World War II order, when viewed on a grand scale, achieved many of the objectives of earlier systems. For nearly fifty years, the balance of power, or bipolarity, that marked its essence bounded norms of acceptable behavior, discouraged great power conflict, and brought a multidimensional focus to global stability. A complex set of agreements, relationships, treaties, and supranational bodies sought to address the economic, political, and security challenges confronting a changing world. From the Marshall Plan to the United Nations, initiatives that reflected both realpolitik and traditional stabilizing concepts were created on both sides of the East/West divide. In Haass's view, the United Nations, specifically, reinforced the Westphalian concepts created centuries earlier. State sovereignty, sovereign equality among all states, and nonintervention in domestic affairs were all codified in the UN Charter.[11] To be sure, there were miscalculations on both sides, as conflicts or confrontations in Korea, Cuba, and Vietnam attest; there was also pressure to adapt the United Nations to the realities of a new world order. Stopping or preventing genocide (Bosnia, Kosovo), redressing territorial aggression (Kuwait, Georgia, Ukraine), responding to terrorist attacks (Afghanistan), and preemptively dealing with perceived threats (Iraq, Iran, North Korea) were all tabled for UN action with, as we now know, mixed results.

The current world operating system, which Haass terms World Order 1.0, served relatively well in the second half of the twentieth century, enabling unprecedented collaboration and economic growth while forestalling great power conflict in a bipolar world.[12] If one accepts the computer analogy, I would argue that, in reality, there have been actual and attempted iterations to the operating system, "patches" in computer terminology, that have reflected the creation and dissolution of international organizations, the formation and abandonment of alliances, and the rise and decline of economies, societies, and military capabilities and the nations they support. There has been an evolutionary improvement in the capabilities and capacity of the global operating system. But it is also true that the system has reached its limit; it cannot keep pace with today's challenges and changes. Some, in other contexts, have called for a simple "reset" of the global operating system, but it is increasingly apparent that that will not be enough. The issue is both capacity and speed. A system designed for diplomacy, deterrence, and mutual dependency and defense among nations does not necessarily have the "bandwidth," capability, or resilience to deal effectively with nonstate actors, criminal syndicates, nongovernmental organizations, transnational businesses, or information warfare from whatever source. Similarly, the processes of the past century are not adequate for the present, where challenges come at light speed, in unprecedented volume, and from disparate, diffuse, and, often, dark sources. No amount of automation or artificial intelligence can substitute for erroneous assumptions or ineffective or outdated processes. At the risk of sounding glib, applying more computing power to a flawed process simply gets you the wrong answer faster.

> *The population of Facebook exceeds the population of the largest nation-state.*   —Niall Ferguson

Problem Statement

Technology is changing our lives and redefining the structure and elements of the global operating system. Rapid developments in artificial intelligence, autonomy, cyberphysical systems, networking and social media, and information (or disinformation) flow are also profoundly altering the global security landscape. Nation-states have new tools at their disposal for political influence while they simultaneously create new vulnerabilities to attacks. Nonstate groups and individuals are empowered by social media and radical transparency. Artificial intelligence and autonomy raise profound legal and ethical questions about the role of humans in conflict and war.

The role technology plays in the current national security context is an equally revolutionary, if not a wholly radical, departure from the past. Not since the development of nuclear weapons has such great technological change affected so much in the national security realm. From the ability to wage war by pinpointing a human target from thousands of miles away with an unmanned aerial drone, to the ability to disrupt a space program with a computer virus, to the ability to genetically engineer in a kitchen a highly virulent pathogen that could kill tens of thousands—these all represent ways in which technology has fundamentally altered the landscape of the national security space. Additionally, as observed by Raymond DuBois, "High-technology weapons are no longer the exclusive domain of only a few nations." Both smaller states and nonstate actors are adopting advanced technologies into their warfare. This "democratization" of technologies of destruction, alongside those for enhanced communication and surveillance, creates "a threat landscape unlike any we have faced."[13]

From a national security perspective, the concept of redefining the global operating system looms large, prompting questions of resources, authority, and accountability. It can also seem more urgent to put exploration of those concepts aside to focus on a growing list of more specific and more easily defined modernization and recapitalization requirements as we attempt, in a fiscally constrained and increasingly threatening world, to define where to put each invested dollar to leverage to best effect its enhancement of our national security. For example, significant potential resides in weaponry technological advancements, often termed a "third offset" strategy, announced by then secretary of defense Chuck Hagel in November 2014. As summarized by Timothy Walton in the July 2016 *Joint Forces Quarterly*: "Secretary Hagel modeled his approach on the First Offset Strategy of the 1950s, in which President Dwight D. Eisenhower countered the Soviet Union's conventional numerical superiority through the buildup of America's nuclear deterrent, and on the Second Offset Strategy of the 1970s, in which Secretary of Defense Harold Brown shepherded the development of precision-guided munitions, stealth, and intelligence, surveillance, and reconnaissance (ISR) systems to counter the numerical superiority and improving technical capability of Warsaw Pact forces along the Central Front in Europe."[14]

The potentially significant costs, long timelines, technological uncertainties, and rapidly evolving adversaries all complicate our planning for the future, even as we wrestle with the realities of the present. A decade and a half of conflict has left the United States struggling with the cost of recapitalizing air, land, and sea forces ridden hard over many years: achieving the right balance of technologically innovative and classic manpower-intensive capabilities, of conventional and special operations forces, and the potential and limitations of technology across a growing number of domains. Today's decision-makers understand that things are changing. But they cannot yet discern whether they are on a linear track to a wholly new national security environment or at the cusp of a dimly recognizable cycle that returns us to a more technologically advanced version of a world we once knew of peer competitors, increasing confrontation, and, if not a Cold War, at least a Hot Peace.

A Way Forward

But even as one is drawn to these budget details and procurement pro-grammatics that will, inevitably, shape the global operating system and national security readiness for good or ill, there are even more funda-mental questions that need to be addressed. It is not my intent to specify all the elements of what could or should come next: World Order 2.0. Instead, I will pose questions and postulate issues that will need to be a part of the effort, not just on the part of the national security enterprise or on the oft-cited "whole of government" approach, but, I hope, appli-cable to the "whole of nation" and, indeed, global effort that will be required. While details must quickly follow, the fundamentals of mili-tary and national security planning still apply, as does George Shultz's aphorism to focus on things that can be done and not merely admire the problem. The first question should be: "What are we trying to accom-plish?" followed by the corollaries: "With whom, where, when, and (per-haps most importantly) why?"

Importantly, when we hear the term "security," images of the Depart-ment of Defense and the uniformed services and supporting three-letter agencies come to mind. But defense of the global operating system now demands a much broader context and is the responsibility of a much more diverse group. Today's definitions of "national security" and "the global operating system" encompass economic, political, diplomatic, informational, humanitarian, and educational elements. It is no longer just the role of those who wear the "cloth of the nation." If you are read-ing this, you are or should be a part of the effort.

What?

The national security planning process for operations or contingencies begins with a National Security Strategy. Provided by the nation's senior civilian leadership, it addresses the national interests, goals, and priori-ties, while integrating all elements of national power and reflecting all extant national security directives. Somewhat confusingly, a National

Defense Strategy, then a National Military Strategy, and, finally, Joint Operations Concepts follow it. This bureaucratic process is often agonizingly slow, and consensus among all involved is difficult to attain. But, when complete, it can provide a context or template against which all national security efforts can be measured and a means of ensuring consistency and coherence in answering the most fundamental of questions: "What are we trying to accomplish?"

The difficulty of this process is well understood by my distinguished former colleague General Jim Mattis. He is famously quoted describing Washington, DC, as a "strategy-free zone." Somewhat ironically, in his current position as secretary of defense, he is now accountable, in part, for that strategy's creation. He is as capable as anyone of that task, but in recent testimony in front of the Senate Armed Services Committee, even he noted: "We entered a strategy-free environment, and we are scrambling to put one together." He continued, "Anyone who thinks an interagency, whole-of-government strategy can be done rapidly is probably someone who hasn't dealt with it." But a strategy for preserving national security and defending the global operating system is essential. We cannot expect that the hundreds of elements contributing to national security can act in concert, absent an overarching concept and consistent, coherent goals. No matter what the national security issue, be it terrorism, cyberthreats, immigration, trade, China, Russia, North Korea, or diplomatic initiatives and alliance commitments, each must be confidently and dependably addressed to avoid confusing friends and enabling foes. We also cannot afford to lurch from crisis to crisis, dealing tactically with what, inevitably, will become problems with strategic dimensions. Tactical energy in a strategic vacuum is a recipe for disaster.

As noted earlier, America's technological revolution has transformed our own national security capabilities, an essential part of deterring and defending against attacks on the global operating system. Our military forces and, indeed, those of allies and adversaries were eager and early adopters and now have capabilities previously unimaginable in every corner and at every level of the battlefield. Terms such as "network-centric warfare" and "information dominance" have entered the lexicon and, in some cases, departed as the realities of both the traditional nature of combat and the growing capabilities of adversaries have closed the digital divide. In the domain of cyberwarfare, the creation of more "information nodes," a euphemism for participants, also creates more potential targets and vulnerabilities; unsurprisingly, robustness and resiliency still matter. This is particularly true at the high end of our military capabilities, such as global communications and sensors, major platforms, and, of course, our nuclear deterrent forces. The newer systems need additional hardening against the cyberthreat; ironically, the age of some elements of the nuclear command-and-control system makes them less vulnerable to today's electronic probes and postulated future attacks. As we recapitalize these forces, robust, flexible, and adaptable cybersecurity elements must be designed in, not bolted on.

It is also true that, driven by the pace of change in the character of the threat, our approach to the development, procurement, and deployment of resilient and adaptive systems must change. It has almost become an article of faith that the government procurement process is an antiquated and bureaucratic process. In the words of defense analyst Loren Thompson, it "is as baroque as it is broke." Study after study has echoed the Packard Commission's conclusion in 1981 that there was "no rational system" governing defense procurement and that it was not fraud and abuse that led to massive over-expenditures, but rather "the truly costly problems are those of overcomplicated organization and rigid procedure."[15] And, finally, we must get it all done faster simply because the threats to the global operating system are advancing faster than our ability to counter them.

To be sure, there is work being done within the Department of Defense. Initiatives such as the Defense Innovation Unit Experimental (DIUx) and the Air Force Rapid Capabilities Office (RCO) are a start but lack the scale and funding to broadly reshape the procurement process. Key elements that must be included, according to former assistant secretary of the Air Force William LaPlante in 2015 Senate testimony, are "strategic agility and adaptability principles" aimed at fielding resilient systems more rapidly—resilient in the sense that they "are inherently resistant to predictive failure." In particular, LaPlante stressed the use of modular designs, open architectures, and "block upgrades" to shorten development cycle times, enable continuing competition, and keep pace with dynamic threats. But what the DoD and the services can do is limited; they are still captive to rigid budgeting cycles, focused congressional oversight, and thousands of sometimes contradictory laws and regulations. An old friend, retired Air Force general Joe Ralston, is fond of saying, "You know when you're in a real crisis because that's when they suspend all the rules!" In many areas of national security, we are there—and it is time to appropriately unleash our world-class developmental, manufacturing, and operational expertise. We cannot and must not wait for the existential crisis to grant us the capabilities we need.

On the practical side, the new capabilities we do have are legion; near real-time imagery, electronic intelligence, drones, nascent directed-energy weapons, and offensive cyberwarfare are but a few. While I often argue that change itself is not hard, the pace and, in this case, the acceleration of change is creating its own set of complications. The resultant challenges include high skill demands on the part of our forces, the lack of precision in cyberattacks, the creation of scarce high-demand, low-density resources, and the need for "exquisite" (near-perfect) intelligence for their effective employment. The speed of our technological advances across specific military programs has introduced or exacerbated the real problem of inadequate communication between our systems, among our services, and, importantly, with our allies. In the case of new technologies or confrontational domains, such as space or cyberspace, our policies and legal or ethical concepts have struggled to keep pace. What does deterrence look like in space and cyberspace, for example, or the concepts of proportionality and discrimination, long a part of the law of armed conflict? Simply extrapolating terrestrial kinetic concepts is patently insufficient.

The military is often accused of preparing for the last war when, in fact, it is the military that is expected to simultaneously "learn from history," deal effectively with today's challenges, and perfectly predict and respond to the future. Ensuring the nation's security and the protection of the global operating system that supports it is a capstone exercise in dealing with risk. In our resource-constrained, threat-rich environment, we simply cannot do it all or expect perfection in every one of those tasks we choose to undertake. Prioritizing the risks that inevitably confront us and deciding, specifically, both what we will and will not do is an essential first step. Nothing of any consequence we do as individuals, as nations, or as a global community is ever risk-free.

Our challenge is to pursue success in each of what I call the Four M's: *measure* risk, *minimize* the risk to the extent possible, *manage* the risk that inevitably remains, and, finally, be prepared with a *mitigation* plan when the next crisis materializes.

Measuring Risk

Measurement of risk is not and has never been easy. The ability to use past or present data to predict future events can be plagued by insufficient data or, as is likely in these days of so-called big data, overwhelmed by far more than we can possibly assimilate. In the national security context, the fog of war has gone digital. Another pitfall is that we analyze the wrong data set or rely excessively on standard metrics or indicators that may not be relevant to our real needs.

Those defining the "what" in defending the global operating system must also understand the wisdom of Pascal's Wager, which reminds us, as we prioritize our many goals and objectives, that the *probability* of an event is not the same as the *consequences* of an event. That is why discussion of nuclear deterrence must still bookend the national security conversation that then flows across multidomain conventional conflict to unconventional warfare and, now, potential confrontations in space and cyberspace.

Modern disruptive technologies can't in general be compared to the wholesale massive destruction of nuclear war. But in some cases, I think there's evidence that nations could be brought to their knees, or societies could be brought to their knees, if attacked by these new technologies.   —Raymond Jeanloz

---

Minimizing Risk

Measuring risk helps with the next step: minimizing it. The national security domain struggles with discerning the difference between large and small risks and understanding where influence can be most effectively leveraged. At a European strategy seminar some years ago, I sat next to the CEO of a British aerospace firm. During a break, I asked him the key to his corporation's recent success. Somewhat simplistically he replied: "Jim, it's simple. I hire the best people I can get and give them everything they want to succeed. And then I let them do it." Unfortunately, I never got to ask him how he could tell the difference between what they wanted and what they needed, as it seemed to me a question any fiscally constrained CEO would want answered.

And so it is with minimizing risk: the possibilities are limitless, but the resources, whether fiscal or personal attention, are not. So where should we place the focus? Minimizing possible risks in new "operating system" designs is important, but so are reliability and maintainability enhancements to existing elements. Hiring and retaining of quality personnel is one factor, but so is wrestling with the knowledge transfer issues surrounding an aging workforce, declining manpower, and blazingly new technology.

Managing Risk

After assessing the risk with as much definition and fidelity as possible, and then working to reduce it to the lowest levels possible, it is surprising how often organizations assume that what happens next is somehow beyond their control. But risk acceptance is not risk management. This is particularly true with technologies that one may be able to fully employ while only dimly understanding the engineering or electronics that make it all work. A few years ago in Europe I came across an old German military maxim expressing a similar abject surrender of control. It says: "All skill is for naught if an angel wets the flintlock of your musket."

An opposite reaction among those responsible for global security is to strive to constantly and actively manage risk to ever-lower levels. But defending the global operating system does not require one to become a systems engineer, only that one understands fully the capabilities and consequences of one's technology. It requires leading and holding accountable a capable team and cultivating independent sources to confirm assumptions and monitor progress.

As Nobel laureate Daniel Kahneman has observed, effective risk management also requires the courage to trust personal instincts about things that just don't sound right. He describes in his book, *Thinking Fast and Slow*, his creation of a standardized screening process for candidates to join the Israeli military. Using a series of factual questions, on a one-to-five scale that focused on six specific traits, he initially eliminated the intuition of the interviewers as a factor, believing a numerical assessment had more consistent validity. When the interviewers objected to his "turning us into robots," Kahneman reluctantly added a final step. He added the requirement for the interviewer, when all earlier steps had been meticulously completed, to "close your eyes, try to imagine the recruit as a soldier, and assign him a score on a scale of 1 to 5."[16] After several hundred interviews and performance feedback from their commanding officers, the new interview process using the six factors proved to be a dramatic improvement in predicting a soldier's success. The big surprise was that the seventh element, the "close your eyes" exercise, did just as well. From this, Kahneman concluded that "intuition adds value . . . but only after a disciplined collection of objective information and a disciplined scoring of separate traits. . . . A more general lesson . . . was do not simply trust intuitive judgment— your own or that of others—but do not dismiss it, either." Instincts should never be the only rationale for critical decisions, but they can help alert when to ask more questions, decline to accept conventional wisdom, seek a second opinion, or move with caution when "something just doesn't seem right."

The very nature of leadership is you better decide before you can know. Those steeped in the humanities are conditioned to be relatively more comfortable and able to handle that.   —Charles Hill

---

Mitigating Risk

Risk mitigation is the fourth and, in some ways, the most challenging of the "Four M's." When I lecture in a class on risk analysis here on the Stanford campus, I take great pains to point out that only a small portion of risk mitigation occurs *after* an untoward event. The bulk of the effort involves preparation in both systems design and resiliency and emergency response capability, coupled with thorough training and regular exercises.

Over recent years, roiled by the events in our country and around the world, we have seen our daily focus changed dramatically, driven, in some cases, by perceived attacks on our global operating system from cyberthreats, so-called fake news, continuing terrorist attacks, and political churn at home and abroad. National security professionals have reacted by seeking information, providing assistance, reassuring stakeholders, and beginning to shape a response. Meanwhile our personal reactions, no matter what our political persuasion, have, successively or simultaneously, probably included concern, disappointment, defensiveness, and even anger. But the most thoughtful have certainly paused, stepping back from the press of today's crises, and considered not just what we are as a nation but what we might and must become after all of this is done.

A principal area in which I believe there are risk mitigation lessons to be learned is in emergency response. It is true that our nation has had a local or regional emergency response obligation and capability for decades requiring emergency plans, emergency response centers, and coordinating or controlling organizations in the case of natural disasters or terrorist attack. Similarly, at the highest levels of the national security apparatus, we have gathered in the "Tank" at the Pentagon or as the National Security Council to address specific international or national security challenges. But how will we respond—and who will—to the large-scale, multidimensional crises that may still confront us?

For example, all of us past our twenties remember the horrific events of September 11, 2001. We remember where we were, how we learned of it, how it changed our lives forever, and some of us still mourn the friends and colleagues we lost. We look back with pride on the way the nation came together, leadership reassurance, and the outpouring of assistance to those affected as our critical transportation infrastructure ground to a halt. We also remember that, despite all of those successes, few would dispute that it was, at best, a pickup game. The chaotic events of those days have clearly shown the benefits of continuing to significantly improve specific plans while moving beyond them to establishing and formalizing a national and international response capability worthy of the name. I wonder, as those events drift aft in our wake, whether we are today as committed and capable as we might someday need to be.

To be clear, I am not talking about merely rewriting plans, creating national or international working groups, and constructing memoranda of agreement. Rather, I am suggesting that we need to consider having, at the ready, a robust, highly capable response team with pre-delegated authority and pre-staged equipment, interoperable both domestically and internationally. This national emergency response organization—which I, tongue in cheek, call NERO, after the Roman emperor who famously fiddled while Rome burned—could be a powerful and, I believe, collaborative effort in which the nation could visibly take a leading role both domestically and internationally, across a broad spectrum of challenges to both our national and global operating systems.

Two decades ago, an element of the Department of Defense coined the term "virtual presence." Then just on the technological cusp of today's ubiquitous capabilities, the concept got a little traction until a clever low-level officer noted "virtual presence means real absence!" That was the end of that concept. In today's technologically networked world, replete with handheld global communication devices, virtual reality, high-definition videoconferencing, and artificial intelligence, virtual presence is now here and must be a key element of a NERO structure that unites the real national "first responders" to disruptions of the global operating system. Regular interactions, comprehensive policy discussions, sharing of best practices, proactive contingency planning, and regular exercises are but a few of the roles for NERO. Technology will allow true national experts who have important "day jobs" to interact regularly and effectively with colleagues around the world without the need to be continuously huddled together in a basement command center behind a sign that reads "Break Glass in Case of Emergency."

With Whom?

An adaptation of a writing of the rabbinic sage Hillel the Elder is often quoted as: "If not me, then who?"[17] For the decades since the end of World War II, when the question was asked in the West, the answer has most often been: "The United States of America." Through the critical days of the Cold War and beyond, in crises of security, economy, or humanity, we were there. On NBC's *Today Show* (February 19, 1998) and repeatedly since, then secretary of state Madeleine Albright famously declared: "We are America; we are the indispensable nation. We stand tall and we see further than other countries into the future, and we see the danger here to all of us." Nearly twenty years later, even if the statement is still true, as I believe it is, there is more than a hint of arrogance and certainly an implied perfection that, in truth, has not been borne out consistently through the incredible range of challenge and change we collectively confronted over the last two decades. Nevertheless, despite the closing of influence gaps and the rise of a new near-peer competitor, America remains the preeminent global power.

Last year, I was privileged to moderate a session of a Hoover Institution seminar on American exceptionalism with George Shultz and the late Sid Drell. In my brief remarks, I opined that *exceptionalism*, which has often been an American hallmark, was not the same as *triumphalism*, which must never be. Returning to Tim Kaine's recent critique of American doctrine and strategy, "Instead of proclaiming its own indispensability, the United States should strive to reestablish its position as the *exemplary* democracy [emphasis added]."[18] No matter where you are in the spectrum of self-defining America's place in the world, one fact remains incontrovertible, if not unarguable: there are things only America can do and leadership roles only America can play.

That is not to say we must or should do it all or go it alone. To achieve and sustain an effective, resilient, and just global operating system across the world, we must establish, by personal commitment and example, not fiat or decree, global standards with international accountability. Believe me when I say that I do not believe it is our job to change the world; I do believe, however, that it is appropriate to support the world's efforts to change itself. Continuing attacks on the global operating system are international events and demand an international response. While the former statement may be self-evident, the latter is not, and I believe that they are inextricably linked.

Let me explain. In the eyes of many, including many of us, historic disruptions of the global operating system laid bare some significant gaps in our performance and effectiveness as a global community. These events, I hope, have swept away reservations of any who thought that events half a world away could not have significant influence on our domestic security, as we now broadly define it. Similarly, to succeed in both countering and containing the threats, any response we craft must have an international dimension. To do less would be, at best, shortsighted and, at worst, sadly ineffective.

My second takeaway from both my own government service and our national crisis response experience over many decades is the value of relationships, both those of long standing, which are often deepened and strengthened, and those that are created afresh with organizations and individuals who share our concern and commitment but with whom we had never before spoken. Each of these stakeholders is involved in different ways for different reasons, bringing specialized and necessary expertise, skills, and resources that, in concert, can provide essential aid to those attacked and information and insight to us all. My point here is to highlight that those who will come together in any crisis of the global operating system may be united by a commitment but not by process, training, or previous interaction. We should not be reduced to creating new relationships on the fly or introducing ourselves for the first time during initial organizational conference calls.

I often speak of silos in organizations, or, as former national intelligence director Admiral Mike McConnell, tongue firmly in cheek, used to call them, "cylinders of excellence." Going forward, we should look at these new relationships with purpose and a strategic objective in mind. The long-term goal should be a process and a structure that cross those industry, government, and private-sector boundaries to enable consultation and collaboration in time of crisis. Our collective efforts in responding to earlier crises have demonstrated what we each can bring to the table. As a nation and as a world, we need to become better at it.

Advances in AI will make extraordinarily more complicated coalition issues, not just because of governance, but because of speed at which decisions have to be made, or prior delegation in order to train the algorithms to do it. How do we keep a sense of consultative decision-making in this world?   —Kori Schake

---

Where?

Throughout our history we have been singularly unsuccessful in predicting where geographic challenges to our national security will arise. Despite modern intelligence technology, we failed to anticipate events in the Balkans, were surprised by the invasion of Kuwait, did not foresee the scope or pace of the Chinese buildup in the South China Sea, and could not conceive of a scenario in which Russia would annex Crimea. The politically and geographically disparate character of the challenges to the global operating system should remind us of several things. First, we need to be better at seeing the world through the eyes of others, be they friend or "other," and not as predisposed to mirror-imaging. Second, our forces, especially land and maritime, need to be regionally present if we are to shape events before they occur, the essence of deterrence. This presence must be balanced by diplomatic representation, humanitarian assistance, and, here, a supportive presence that must be real, not virtual. You cannot surge trust. Third, we need to appreciate that in every case of applying "bleeding edge" technology, we have erroneously assumed that we are thus operating in a secure sanctuary and that, since adversaries do not exist, they never will. We find ourselves then playing catch-up when confronted by inevitably emerging threats, which we are ill equipped to counter or deter. And finally, we need to understand that we will be appropriately sharing the global security burden with others, each of whom brings unique capabilities, insights, and regional security expertise.

We cannot ignore and must not dismiss the international dimensions of our efforts. I believe passionately in the global security community's obligation to work collectively toward a common goal and, in a previous life, spent decades with valued colleagues from around the world contributing, I hope, in meaningful ways to that effort. To sustain and protect the global operating system from challenges around the world, we must help establish and sustain, by personal commitment, not fiat or decree, global standards of behavior with attendant international accountability.

But there are now new and very different "geographies" to consider, one distant and one omnipresent, that have unique vulnerabilities and on whose systems we are increasingly reliant. The first is the space domain, once considered a remote sanctuary and now increasingly accessible, globally essential, and uniquely vulnerable. The second, of course, is the cyberdomain, which, after more than four decades, has literally transformed our world.

As I wrote in the introduction to a 2016 National Academy of Sciences study:

> The national security of the United States is inextricably linked to space and our unimpeded access to the capabilities resident in or traveling through that domain. Since the dawn of the Space Age, all those who have been a part of what was once a race between two superpowers and is now a $315 billion global enterprise, have implicitly understood this linkage. Over, now, six decades, that reliance on space systems has deepened and broadened. What was once only a realm of exploration and national security has grown to include a commercial element that has become so ubiquitous that it has led us to fundamentally redefine the term "national security space."[19]

One overarching conclusion of that study relates directly to the vulnerabilities of the global operating system in that the speed of advances in access and space-borne capabilities has significantly outpaced the creation of guiding national—let alone international—strategies and policies. The technological advances in space systems and increased reliance on them have created a space-enabled "critical infrastructure" that has not been matched by coherent supporting protection and loss-mitigation strategies, clearly articulated and accepted policies, and robust defensive capabilities.

These concerns are even more relevant to the internet and the ubiquitous prefix "cyber" attached to dozens of terms, from "space" and "commerce" to "security" and "warfare." Indeed, when the term "global operating system" is used, many think only of the hyperconnected world in which we live and the incredibly complex linkage of the internet, with all its capabilities, possibilities, and, yes, vulnerabilities. In introducing one of several cybersecurity initiatives in 2013, the Obama administration noted:

Cyberspace touches nearly every part of our daily lives. It's the broadband networks beneath us and the wireless signals around us, the local networks in our schools and hospitals and businesses, and the massive grids that power our nation. It's the classified military and intelligence networks that keep us safe, and the World Wide Web that has made us more interconnected than at any time in human history. We must secure our cyberspace to ensure that we can continue to grow the nation's economy and protect our way of life.[20]

It's one thing to deal with Russia, where we have a long history of interactions and joint recognition of the challenges and so forth. How do we put this in the context of a world with nonstate actors and, particularly, someone who is either crazy like a fox or an absolute madman? It strikes me that we've got a whole new set of challenges here.   —Thomas F. Stephenson

---

When?

In short, we can wait no longer. The efforts at collective solutions to global problems often identify the building of a global consensus on principles, details, and implementation as the most daunting challenge. Regrettably, I believe that the term "global consensus" is increasingly an oxymoron and, even if ultimately achievable, will come at a pace that we all know is too slow to satisfy both our needs and the security expectations of our nations. I also hope that we are not waiting for or proposing the creation of yet another organization, clearinghouse, or global coordinating body. The need is not for more structure or nonproductive bureaucracy; it is for more effective and collaborative use of what we have. This is not the time to engage in long-term discussions on roles and responsibilities or for dramatic shifts in oversight scope or accountability in an Al Haig-like effort to declare "I'm in charge!"

We must expand and enhance the collaborative efforts of which this symposium is a fine example and drive real change in outcomes, not organization, content in the belief that, if organizational changes are necessary down the road, the form should follow function. If we get the "what, when, and why" right, the how will follow. Organizationally, I often note how sidewalks should be placed on a college campus: where the paths are worn in the grass. That is the clear indicator of how interaction really works, in practice, not in theory. We need not and should not try to define that structure first. Again, there is important work to be done now.

For many years, in the much different context of commercial nuclear safety, a few others and I spoke passionately of establishing a "coalition of the willing" and the potential that it represented. I now have come to believe that I, at least, was thinking too narrowly. What is really necessary to deal quickly and effectively with emergent challenges to the global operating system is a coalition of the ready, willing, and able. *Ready* to gird for the battle now, not at some future moving milestone; *willing* to act, not discuss, debate, or delay; and *able* to bring real resources, drive real change, and demand real accountability.

I am not suggesting we abandon those that are not ready for the journey. But we cannot wait for them to prepare fully. We cannot and should not wait to find what some call the "common denominators," which can, if we are not careful and as the mathematicians reading this know, also often include the modifier "lowest." And, finally, we cannot let them slow the pace.

For years, in a previous life, I spoke of the differences between an alliance and a coalition. An alliance has a formal structure, demands unanimity, and often requires extensive debate and concessions to a myriad of partner concerns before acting. The positive aspect in an alliance, of course, is that when it ultimately ceases talking and moves to action, it brings everyone to the task with the strength of numbers and unity of purpose. A coalition, on the other hand, is like the planting of a flag in the ground. A common goal, a shared objective, and full agreement are demonstrated simply by the participation of those who voluntarily rally around the standard. The only metrics are immediate action and real achievement in pursuit of time-critical goals.

Whether we are comfortable discussing it or not, our world, like our societies, is composed of nations and organizations of diverse skills and wide-ranging, variable capabilities. Those that are stronger in expertise or experience, resources or resolve can seize and shape opportunities out of crises that daunt and discourage others. Thomas Carlyle once said, "The block of granite that was an obstacle in the pathway of the weak, became a stepping stone in the pathway of the strong." The image of which I write today is this: those that *can* must *do*, those that are *able* must *achieve*, and they must *lead* so that others may *follow*. They must form or formalize a coalition that is an example and a standard for all.

What keeps gnawing at me is the speed.   —George P. Shultz

Why?

When I began this chapter, I thought that this, the final section, would be the longest and most nuanced. I was wrong. What has gone before has convinced me, at least, of a few fundamental truths. First that the global operating system, as I have defined it, is under constant attack and the threat is growing geometrically. Second, I believe that our ability to deal effectively with the diverse and diffuse challenges has declined, even as the importance of doing so has increased dramatically. Finally, I believe that only the United States can collaboratively and collectively lead this effort; only we have the resources, the global role, and the resolve to get it done.

There are some who believe that, weighed down by the burdens of the last decade and a half, the nation has tired of global leadership and that isolationist sentiments are on the rise, fed by populist trends and a growing sense that national priorities lie elsewhere. This is not the first such conversation to take place in the United States. An earlier version occurred in 1947 when President Truman advanced the Marshall Plan and a vision of America's essential leadership role on the cusp of the Cold War. He, too, faced a nation strained by a global conflict and national sacrifice, understandably leery of entangling global alliances, much less a confrontation with the Soviets. Truman was able to paint a realistic, if frightening, picture of the Soviet threat to Europe and warn that if Europe fell America would follow. In a recent *Wall Street Journal* op-ed, Walter Russell Meade noted, "A Trumanist approach—popular but not populist, moral but not moralistic—would start by showing some trust in the American people. To take one obvious instance where popular and elite views diverge: Ordinary people are inclined to favor a firm, decisive response to jihadist threats, while foreign-policy elites tend to worry much more about the possible effects of American overreaction."[21]

Today, we need another such candid conversation with the American people, one that, rather than creating hysteria, both increases understanding and inspires confidence. As Henry Kissinger reminded the US Senate in 2015 testimony with Madeleine Albright and George Shultz, "The problem of peace was historically posed by the accumulation of power, the emergence of a potentially dominant country threatening the security of its neighbors. In our period, peace is often threatened by the disintegration of power, the collapse of authority into non-governed spaces spreading violence beyond their borders and their region."[22] That technology has played a pivotal role in this change is unarguable. On its website, the Center for a New American Security, perhaps reflective of the word "New" in its name, notes:

Technology is changing our lives. Rapid developments in artificial intelligence, autonomy, cyber-physical systems, networking and social media, and disinformation are profoundly altering the national security landscape. Nation-states have new tools at their disposal for political influence as well as new vulnerabilities to attacks. Non-state groups and individuals are empowered by social media and radical transparency. Artificial intelligence and automation raise profound questions about the role of humans in conflict and war.[23]

Conclusion

I began with a discussion of the historical context of world order and my corollary for today's world, the global operating system. I have attempted to outline the scope of the challenge to today's system in the context of the failures of those past. There are lessons to be learned. But they are not lessons learned merely because we write them down; something needs to change, to be approached and dealt with differently. These times are very different and in so many ways more fraught with ubiquitous risks and threats, some unfolding at light speed in nonkinetic but equally impactful ways. It is also too easy, and patently incorrect, to demonize the recently emergent and exponentially exploding technologies. Each problem or challenge attributed to today's technologies is mirror-imaged by many more capabilities and benefits that can improve the lives of tens of millions and, in so doing, enhance our global humanity. The growing challenge we face, as Christian Lange cautioned, is finding a way to remain the master of it all and not become its servant, much less its victim. This is a human challenge, not a technological one.

To be sure, the technological and policy debate, followed by real and substantive assessments of the way forward, will demand an unprecedented level of candor and, certainly, confrontation among all participants. Legacy platforms and processes, not to mention policies, must be rigorously examined and questions of current effectiveness and future relevance honestly addressed. The real resource challenge of system replacement must be balanced with the realities of mitigation effectiveness. In the space domain, for example, considering the potential for widespread GPS jamming or an on-orbit electromagnetic pulse attack, we are already hearing calls for a return to a pre-GPS national security world using updated systems of the past such as E-Loran or thumb drive-size inertial navigation systems. Yet, even as we know the questions to ask, we lack even the analytic tools to dispassionately quantify the operational and fiscal costs that must be a part of the answer as we wrestle with the viability of balancing "the way we have always done it" with the costs and uncertainties of technologies yet to be defined. The old naval maxim comes to mind: "Never let go of one rope until you have a firm grip on another." We know instinctively that Abraham Lincoln was right when, in his annual address to Congress in 1826, he said: "The dogmas of the quiet past, are inadequate to the stormy present. The occasion is piled high with difficulty, and we must rise with the occasion. As our case is new, so we must think anew, and act anew." But, as we all instinctively know, addressing the difficult realities of defending the global operating system is not, at its heart, a technical issue. It is a leadership challenge.

Organizational management scholar Edgar Schein has written that one of the primary roles of leaders in time of crisis is to absorb fear, not create it, through clear communication, a demonstrated understanding of the problem, and swift, inclusive action to deal with the looming realities. In essence he defines what Jim Collins calls the "Stockdale Paradox" after the storied Vietnam prisoner of war and former Hoover fellow. Admiral Jim Stockdale told him: "This is a very important lesson. You must never confuse faith that you will prevail in the end—which you can never afford to lose—with the discipline to confront the most brutal facts of your current reality, whatever that might be."[24]

And the facts, as I have attempted to describe, are brutal. A recidivist Russia, a forcefully rising China, a capably belligerent North Korea, and a virulent, violent strain of Islam intentionally confront an increasingly fearful and uncertain global community. Robert Kagan, in an article ominously titled "Backing into World War III," writes: "Americans tend to take the fundamental stability of the international order for granted, even while complaining about the burden the United States carries in preserving that stability. History shows that world orders do collapse, however, and when they do it is often unexpected, rapid, and violent."[25] He brings our meditation full circle when he goes on to note:

For the United States to accept a return to spheres of influence would not calm the international waters. It would merely return the world to the condition it was in at the end of the 19th century, with competing great powers clashing over inevitably intersecting and overlapping spheres. These unsettled, disordered conditions produced the fertile ground for the two destructive world wars of the first half of the 20th century.[26]

I end with a bit more optimistic thought from Secretary Shultz, who, like Admiral Stockdale, has the knack of balancing realism and optimism. In his book *Issues on My Mind*, he reprises remarks he delivered to the Commonwealth Club of California in 1985: "Civilizations decline when they stop believing in themselves; ours has thrived because we have never lost our conviction that our values are worth defending. But America also has a moral responsibility. The lesson of the postwar era is that America must be the leader of the free world; there is no one else to take our place."[27]

It is fitting that I also include a final thought that might have been conveyed by Sid Drell, were he with us here today—and who's to say he's not? I quote from a copy of a book he authored with McGeorge Bundy and Bill Crowe in 1993:

One of the great lessons of the last few years is that change is sometimes fast and large and good—and also unexpected. What can be said for now is that both hope and danger make this an extraordinarily good time for continued effort. That effort cannot be American alone, but it cannot be much without us.[28]

In a final note, inscribed on the flyleaf of the copy of Sid's book from which I quoted, are the words: "To George Shultz, with warm friendship, Sid Drell."

5

TECHNOLOGICAL CHANGE AND GLOBAL BIOLOGICAL DISEQUILIBRIUM

Lucy Shapiro and Harley McAdams

We are living in a time of explosive manmade change. Breakthrough advances in the sequencing, decoding, and manipulation of genomes of all organisms are occurring at the same time as disruptive changes in the world's ecosystem. We are in the midst of the sixth great extinction, which is predicted to culminate in the elimination of about 30 percent of all ocean corals, sharks, and rays, 30 percent of all freshwater mollusks, 25 percent of all mammals, 20 percent of all reptiles, and about 15 percent of all birds currently alive.[1] Many factors contribute to this global disruption, including increasing carbon dioxide (CO_2) leading to climate change and shifts in ocean chemistry, geopolitical upheavals, poorly controlled intercontinental transport, unsustainable population growth, deforestation, and urbanization. Human actions are behind these factors that are eroding our ecosystem, and it remains to be seen if the coincident advances in technology can mitigate the consequences of the mess we are making.

These disruptions also are having direct consequences on human health. Climate change is a cause of the global redistribution of infectious diseases. Modern mobility facilitates rapid spreading of new diseases or pathogens that emerge from remote corners of the world or by mutation of familiar diseases. Excessive or inappropriate use of antibiotics anywhere can have dire medical consequences globally. While many medical professionals and biological scientists have become acutely aware of these concerns over the last two decades, the principal responses have been largely reactive. Examples include emerging disease surveillance and response systems, improved clinical hygiene, and international coordination through the World Health Organization (WHO). But there is severe underinvestment in some critical areas. Two examples are surge capacity for global vaccine production to respond to a severe influenza pandemic and investment in basic research and drug development to respond to the ongoing and inevitable decline in effectiveness of current antibiotics.

The Global Redistribution of Infectious Diseases

Even small, one- to two-degree Celsius changes in ambient temperature can alter the habitat and thus the global distribution of viral, fungal, and bacterial pathogens and the birds, mice, ticks, rats, bats, and mosquitoes that carry them. Between December 2014 and June 2015, nearly fifty million domestic poultry in twenty-one American states were slaughtered to stem a raging Asian avian flu contagion. This was the worst animal disease pandemic in US history. How did it happen? Global warming has shifted migratory bird flight paths, leading to an overlap of the south-to-north Asian Pacific flyway and the North American Pacific flyway at the Bering Strait. The Arctic waters are warming faster than other regions on earth so that the Bering Strait has become a meeting and mingling spot for flocks following flyways that formerly rarely mixed. DNA sequencing enabled identification of specific Asian avian flu strains that were hitching a ride in these mingling flocks as well as their sites of origin and their mutation rates. In late 2014, an Asian avian flu virus that was transferred to the North American flock appeared first in Canada, followed by Oregon, Idaho, and Washington State. We were fortunate that the avian flu strains tracked during this period did not easily infect humans or transfer between humans. However, with continued mutation of the virus and re-assorting of genetic material among mixed viral populations, the generation of a human pandemic strain of flu can happen at any time. The question is not if, but when such a strain will arise.

The organisms of the world that have evolved over millennia are adapted to thrive in local or regional ecosystems. Now, however, we are experiencing rapid global movements of formerly local pathogens and their vectors. These changes affect the health of people, ocean life, and the animals and plants that are our food sources. The living world is in trouble!

Since the 1980s, the annual number of epidemics across the globe has tripled, leading to social and economic disruptions. Fungal infections of corals weakened by warming and more acidic oceans have decimated the coral reefs that are part of the foundation of the ocean food chain. Hundreds of millions of people worldwide depend on them for their food and livelihoods. The death of reefs, from the Australian Great Barrier Reef to the reefs in the Caribbean, is causing a catastrophic disruption in the global food chain.

In addition to humans and animals, plants fall prey to epidemics. The precipitous rise in infections of food plants, such as fungal infection of the world's banana crop and bacterial threats to citrus crops, can cause global disruption of the food supply. And we are forgetting the lessons of the past. The devastating potato famine that killed one million people in Ireland between 1845 and 1852 resulted because of wide dependence on a single strain of a common food staple, the Irish Lumper potato. These potatoes were susceptible to blight caused by the *Phytophthora infestans* fungus that arrived in Ireland in 1844, leading just two years later, in 1846, to loss of three-quarters of the potato harvest to the blight. The unfolding crisis in the commercial banana industry is somewhat similar. Almost all bananas in our stores are the Cavendish variety. Every Cavendish banana plant worldwide is a clone and thus is genetically identical to every other. This is a recipe for disaster as a disease capable of killing one plant can kill them all. There are now fungal diseases that will wipe out the Cavendish banana within a decade. Sadly, this scenario was both inevitable and predictable.

Global climate change contributes to unprecedented exposure of crops, livestock, wildlife, and humans to new viral, bacterial, and fungal pathogens as well as their vectors. This abnormal mingling in the biosphere causes rapid emergence of novel pathogens and the appearance of old pathogens in new places. The changing pattern of wildfowl migrations is just one example where the distribution of global wildlife is being disrupted. Weather pattern changes bring insect vectors that carry viruses into new population centers. We routinely share living quarters with the *Aedes aegypti* mosquito, which has adapted to life in urban areas, doesn't wait till evening to bite us, and carries multiple dangerous viruses. Examples include dengue (also called breakbone fever), chikungunya, yellow fever, and Zika viruses. These are just a few of the viruses we know about! The breeding habits of *A. aegypti* are sensitive to temperature: when ambient temperature increases, their gestation time decreases and their breeding seasons become longer. They like warm weather and standing water, and changing weather patterns are now providing plenty of both.

Counterintuitively, drought conditions can also lead to increases in mosquito-borne infections. An example is West Nile virus. This virus first made landfall in the Northern Hemisphere in 1999, with the sudden appearance of dead birds in New York's Bronx Zoo. The West Nile virus is transmitted by mosquitoes to both birds and humans. During droughts, birds and the water-loving mosquitoes are frequently together at the scarce water containers, leading to increased chance of virus transfer and propagation.

The tropical areas around the middle of the globe are the traditional habitat of the *A. aegypti* mosquito. Over the past several years, concurrent with the dramatic retreat of the Arctic and Antarctic glaciers caused by the warming of the oceans, there has been an equally dramatic change in the geographic distribution of mosquitoes and the pathogens they carry. Because of the temperature-induced migration of the mosquito vector, many disease-causing microbes have moved out of the tropics and into the temperate zones. The establishment of new disease zones is further enhanced by the high mobility of modern human populations. A case in point is dengue fever, with over four hundred million people infected per year in tropical zones worldwide. Dengue is now newly established in the Caribbean and throughout the state of Florida as well as increasing in multiple southern US states and California, coinciding with the new distribution of the *A. aegypti* mosquito. Chikungunya, a mosquito-borne tropical virus, has emerged in South America and the Caribbean, with over a million infected people. Hundreds of chikungunya infections of people who have not been previously exposed to this disease have been reported in the continental United States. Zika virus, another tropical virus spread mostly by the bite of an *Aedes* species mosquito, was first identified in the 1950s in the monkey populations of equatorial Africa. It began its rapid migration a few years ago, first to Polynesia, then to Brazil, Central America, the Caribbean, and finally to the US mainland. Zika can be transmitted by sexual contact or blood transfusions. Zika can also be passed from an infected pregnant woman to her fetus to cause devastating effects on the fetal brain and nervous system.

Currently, there are no effective vaccines or drugs on the market for the dengue, West Nile, chikungunya, or Zika viruses, although international R&D efforts are in place to address these challenges. There are parallel efforts to control the mosquito populations. An effective method, previously used in Brazil and just approved for a small trial run in a southern Florida county, releases mutant male mosquitoes that, when bred with wild-type female mosquitoes, produce nonviable progeny, yielding a 90 percent suppression of this insect vector. Although the method is safe and effective, there has been public resistance to this remedy. Even with advances in vaccine technology, the revolution in gene sequencing and gene editing, and mosquito control measures, staying ahead of the current outbreaks, not to mention those that will inevitably appear, is a significant challenge.

Malaria, which is transmitted among humans by the *Anopheles* mosquito, claims the lives of 650,000 people per year worldwide. Millions more survive with debilitating disease. The dependence of malarial outbreaks on weather conditions is not a new phenomenon. In recent decades, malaria has been primarily restricted to tropical and subtropical environments where outbreaks followed the patterns of rains and floods. Now, however, malarial outbreaks have moved, for the first time, into the highlands of East Africa owing to recent warmer and wetter weather. This has had devastating effects on the newly exposed populations. They have no immunity built up from past exposure and thus are unable to mount an immune response. The result has been a sharp increase in illness and death.

Changes in distribution of mosquito-borne viruses are just one consequence of climate change. Ticks also transmit viral and bacterial pathogens. In the American Midwest, the season for ticks that carry Rocky Mountain spotted fever now starts earlier and ends later. *Ehrlichiosis*, a bacterial pathogen carried by ticks, had been dubbed the "summer flu" because it traditionally appeared only during the warm summer season, while it now occurs any time of the year. In addition, the fungal infectious agent that causes valley fever in California and the Southwest has changed its infectious season. Eleven states in the US and multiple local governments have developed surveillance and containment plans to cope with the spread of the vectors that carry pathogenic agents. We can expect increasing awareness of the geographic change in pathogen distribution as these disease vectors march out of the tropics, and municipalities are faced with providing the funds needed to diagnose, contain, and treat new disease outbreaks.

Headlines announcing the sudden appearance and spread of diseases such as West Nile virus, the viral pneumonia-like SARS (severe acute respiratory syndrome) and MERS (Middle East respiratory syndrome), Ebola, and Zika at new global locations have become an almost yearly occurrence. Climate change is just one of a confluence of factors that have enabled this disruptive migration of infectious agents. Deforestation and population growth push pathogens and their vectors into new environments. Rapid movement of people, goods, and food through porous borders transports the co-traveling pathogens and their vectors.

The twin problems of new infectious diseases and old diseases in new locations are made worse by the growing resistance of pathogens to antibiotics, antivirals, and antifungals. Resistance to the most effective antimalarial agent, artemisinin, is growing rapidly. Owing to the rise in antibiotic resistance, we are now moving precipitously toward a return to the pre-antibiotic era. That would be a medical catastrophe with deaths from infections predicted to exceed cancer deaths and greatly increased risk for even minor surgeries. Again, it is not if, but when.

I do believe we have some evidence on global pandemics that we've responded fairly quickly. My point is, we ought to be having these conversations on a regular basis, whether they're cyber-oriented challenges or whether they're biological or chemical or whatever. I don't think we do that. Every time we have a crisis, it seems as though it's a pickup game.   —James O. Ellis, Jr.

Rising Antibiotic Resistance

Antibiotic resistance is increasing in parallel with the increasing incidence of infectious diseases. US hospitals see two million cases of antibiotic-resistant infections each year that cause one hundred thousand deaths annually. Overuse or improper use of antibiotics is a major factor that can increase the rate of evolution of resistant bacteria. A pathogen can become resistant to an antibiotic in two principal ways: (1) the target organism can evolve a mutation in a gene associated with the structure or function of the antibiotic cellular target; or (2) the bacterium can acquire genes encoding antibiotic resistance mechanisms by importing foreign DNA segments that encode such mechanisms. As one example, bacteria often acquire in this manner a capability to "pump" the antibiotic molecules out of the cell. This bacterial technique for importing DNA is called horizontal gene transfer (HGT). The DNA segments with the resistance mechanisms that are imported by HGT were evolved in other bacterial species that previously encountered the antibiotic in other animals. The resistance mechanism imported by a bacterium infecting a human could have evolved during infection of a different animal by a different bacterial strain. This is how a resistance mechanism that evolved in a bacterium infecting, say, chickens that have been fed antibiotics (a common agricultural practice) can end up in a bacterium that infects humans.

In the quest for economical production, over 70 percent of the antibiotics used annually in the United States are fed to farm animals. This practice inevitably leads to bacterial evolution of resistance mechanisms that can then be passed on to human pathogens as discussed above. There is thus a trade-off between having larger and healthier livestock and the global availability of effective antibiotics. In the short run, we have more meat on the table, but in the long run we will lose the antibiotics that are vital for human health with catastrophic consequences.

HGT is dependent on small circles of DNA called plasmids that carry genes encoding proteins that make a given antibiotic ineffective. Plasmids can be transmitted from one bacterial pathogen species. Thus, when a bacterial pathogen carrying plasmids with resistance genes appears in an environment, these plasmids can spread into all the pathogenic bacterial strains in the area. There are now even "super resistance plasmids" that carry up to fourteen different genes encoding proteins that produce different types of antibiotic resistance. A bacterial pathogen that acquires one of these super resistance plasmids will then be resistant to most antibiotics. A person with an infection of a bacterium carrying such a plasmid is likely doomed.

The good news is that modern medicine can extend human life spans by transplanting organs and even control cancer by means of sophisticated immunotherapy and chemotherapy. The bad news is that use of these medical advances can lead to populations that are immuno-compromised and thus unusually susceptible to infections that must be treated with antibiotics. It is a sad fact that these immune-compromised patients also provide highly effective venues for generation of even more antibiotic-resistant bacteria.

It is important to realize that the recent rapid rise in bacterial resistance to antibiotics is not due to some newly emerged biological phenomenon. Rather, it is a perfectly normal and predictable consequence of widespread unsound clinical and agricultural practices. Clinicians often overprescribe antibiotics, in many countries antibiotics that are available without prescriptions are misused, and the agricultural industry abuses are indefensible. Compounding the problem, for decades funding for basic research in microbiology and training of microbiologists has been in decline due to a misguided belief that the problem of infectious diseases had been "solved" by antibiotics and vaccines. Further, the question of what to do about the rise of antibiotic-resistant bacteria is quite complex. An antibiotic is simply a chemical compound that disrupts some vital biochemical process in the bacterium but is harmless to human biochemistry. But a large majority of the biochemical and genetic mechanisms in human cells are very like the comparable functions in bacteria. We are, after all, descendants of bacterial cells! As a consequence, there are a finite and relatively small number of potential distinct mechanisms that antibiotics can target. The "easy" targets, the low-hanging fruit, were targeted by the antibiotics developed long ago. These older antibiotics are, of course, the ones where antibiotic resistance is more likely to be high since evolutionary selection for resistant bacterial strains is inevitable. We can slow that natural selection process, but we cannot stop it.

The traditional method for finding new antibiotics was to screen literally millions of natural compounds for effectiveness in killing pathogenic bacteria. Those found to have lethal activity were then tested for toxicity in animals and eventually in humans. Unfortunately, this method is no longer productive. The well is running dry. Use of combinations of antibiotics shows some promise for extending the life of older antibiotics. There has also been some success from combining antibiotics with new drugs that attack mechanisms the bacteria have acquired to protect themselves against antibiotics. These are all promising avenues that are effective for some, but by no means all, antibiotic-resistant bacteria.

Each of these new strategies requires R&D funding: basic research funding to discover novel strategies, development funding to reduce them to practice, and clinical trials to demonstrate safety and efficacy. As noted above, basic microbiological research at the federal level has been underfunded for decades. Over the past decade, two-thirds of the antimicrobial research and development programs in big pharmaceutical companies have been downsized, due in large part, but not solely, to the economics of drug discovery. Good medical practice now requires that new and effective antibiotics be kept on the shelf and used only as a last resort. This is to delay the development of resistance. But this practice severely reduces the potential sales and earnings of the new antibiotics that were developed at great expense. Also, antibiotics that are generally taken for just a few days or a few weeks are far less profitable than drugs taken for decades for chronic ailments such as diabetes, heart disease, cancer, and neurological disorders. From the standpoint of business considerations, drug companies' reluctance to invest in development of antibiotic drugs is a rational decision. However, from the standpoint of society's interest in responding to the looming threat of radical increases in the number of deaths from untreatable infections, the situation is insane.

Lucy is telling us that the ultimate arms race is going on right now, and the bugs are ahead.   —Jim Hoagland

---

The Specter of an Influenza Pandemic

Viral influenza, or flu, is an infectious disease of great global concern. Strains of this small RNA virus mutate frequently, changing both the characteristics of the H (hemagglutinin) and N (neuraminidase) proteins that sit on the surface of the spherical virus shell and of the ferocity of the infection. A new flu vaccine is created each year to target the flu strain expected in the coming season. In the 1918 flu pandemic, the

H1N1 strain killed fifty million people worldwide. Subsequent pandemics occurred in 1957 (two million deaths) and 1968 (one million deaths). Not all flu strains are easily transmitted from person to person, but when a mutation in a flu strain causes it to be transmissible, a pandemic can ensue. The story of one flu strain, H5N1 (which is endemic among migratory birds but not currently transmissible among humans), illustrates the hope that biotechnology intervention can enhance viral surveillance, provide a means of containment, and establish effective treatment following an infectious outbreak. It is also a story that raises questions about how, or if, research with infectious agents should be regulated.

The development of technology to sequence DNA and manipulate genomes has been a fundamental breakthrough in the biological sciences. The genome of a virus can now be sequenced in a few hours and the DNA of a bacterium in a day. This technology permits rapid pathogen identification and detailed tracking of disease spreading. The genomes of disease vectors and of humans can also be sequenced accurately and relatively cheaply. Research using these capabilities has produced a much deeper understanding of host-pathogen interactions. Genetic engineering and gene editing have also enabled the design of methods to control the infection process and the viability of pathogen vectors. The databases resulting from this work are the basis of global diagnostics networks for rapid pathogen identification and response intervention. The initial identification and characterization of a new pathogen starts the process of vaccine and drug design and production. Although pathogen identification can be rapid, at least six months of preparation are needed for the design, validation, and ultimately FDA approval for the delivery of a new vaccine. Production and distribution of the quantities needed for a widespread vaccination program take additional months. Development and approval of new antibiotics and antivirals take much longer.

The H5N1 flu strain, first detected in 1997 in Hong Kong, is now carried by poultry and migratory bird populations worldwide. Though there have been just a small number of H5N1 human infections among people in very close contact with infected poultry, the death rate among those who have been infected is over 50 percent. The death rate of the 1918 pandemic flu H1N1 was only 2 percent, but it was wildly contagious, so a vast population was infected. As of now, there has been no known transmission of H5N1 from person to person, but the virus has significant potential to become a transmissible pandemic strain.

Scientists, considering how to respond to this threat before a transmissible human strain develops, are asking three critical questions:

- What is the genetic signature (genome sequence) of a potential pandemic H5N1 strain that can be used for global surveillance?
- Is there a genetic signature that can be used to identify an H5N1 strain with a high kill rate?
- Would a newly evolved pandemic strain be sensitive to existing antivirals and vaccines?

In 2011, experiments in two labs, one in Wisconsin and the other in the Netherlands, addressed these questions by attempting to evolve a laboratory strain of H5N1 that was transmissible from human to human. Their work caused a public conflict relating to the right to perform "knowledge-driven science" versus perceived ethical and security concerns. The ultimate objective of the experimenters was to genetically engineer and evolve a strain of the H5NI flu virus that would be transmissible among ferrets. Why ferrets, and what do they have to do with human transmission? Ferrets were chosen because their response to infection by the flu virus closely resembles the human response. In contrast to infection in birds, infection in both humans and ferrets occurs by inhalation of virus-laden respiratory droplets and subsequent infection by virus attachment to cells in the airways. Then, infected ferrets, like humans, sneeze, spreading potentially infective droplets in their vicinity. The researchers found that it took only five mutations in two genes to generate an airborne transmissible strain of the H5N1 virus. While these mutated viruses are transmitted among ferrets and are sensitive to existing vaccines and drugs, their ability to be transmitted to humans by respiratory droplets is conjecture and not tested.

When researchers from the two laboratories attempted to publish the viral sequence for their potential pandemic strains, the NSABB (National Science Advisory Board for Biosecurity) blocked publication. The NSABB rationale was, first, that there might be an accidental release of the evolved strain and, second, that a published viral sequence might be used to deliberately duplicate the potentially pandemic strain as a biological weapon. A group of prominent scientists called for a moratorium on further H5N1 experiments and for blocking publication of the mutated sequence. This action was reminiscent of calls for a moratorium on applications of genetic engineering at the Asilomar Conference in 1973.

After public discussion and many heated debates, the moratorium was lifted when, in 2012, the US government established an oversight committee that mandated rules limiting the types of experiments that can be conducted with pathogenic strains. Publication of the original H5N1potential pandemic strains was allowed in 2013, and flu research was resumed. Then, in 2014, two things happened. The same lab that had evolved a potentially pandemic H5N1 flu strain succeeded in the reconstruction of the H1N1 1918 flu strain from material obtained from frozen bodies. Second, multiple events involving mishandled pathogens were reported at both the CDC (Centers for Disease Control and Prevention) and at an FDA (Federal Drug Administration) lab.

In response, the Office of Science and Technology Policy and the Department of Health and Human Services mandated a one-year pause in flu research, as well as research on SARS and MERS viruses aimed at eliciting enhanced transmissibility via a respiratory route. Eventually, after two workshops conducted at the National Academy of Sciences and an extensive risk assessment study, the NSABB found that only a small subset of experiments continued to be of concern, including those that "generated a pathogen that is highly transmissible and highly virulent." A plan for oversight of federally funded research in this arena is now in place.

The way that government takes action is through legislation and implementing policy. . . . What feels to me missing from our government though is an articulation of objectives and values that would guide the thinking of people who are in decision-making roles.   —Christopher Stubbs

Yogurt and the Discovery of Gene Editing

As stated earlier, the ability to rapidly and cheaply sequence the genetic material of all organisms has enabled identification and tracking of the migrations of pathogens and their vectors in our increasingly unstable ecosystem. Genetic engineering—the ability to design mutant proteins with altered functions and to move genes from one organism to another—has led to a deeper understanding of how viruses and bacterial pathogens interact with host cells in humans to generate altered modes of pathogen transmission and infectivity. A holy grail of genetic engineering was development of the means to directly edit the genes in chromosomes of all living entities and thus to change the instructions encoded in our DNA. Over the past several years, the ability to edit genes easily and accurately has become a reality. The ultimate impact of gene editing on global health and agriculture is not yet known, but its promise is so far-reaching that understanding how it came about and its potential uses is relevant in any discussion of our technologically driven world. In this regard, the story of the discovery of the CRISPR/Cas9 genome editing technology is a wonderful illustration of the role of serendipity in biological sciences.

The manufacturing process for yogurt uses special bacterial strains to produce the lactic acid that gives yogurt the tang and taste that customers enjoy. But, just as human cells are subject to attack by pathogenic viruses, bacterial cells can be attacked by viruses known as bacteriophage (or phage) that have evolved the ability to attack specific strains of bacteria. Phage infection of the lactic acid bacteria required for the yogurt manufacturing process can bring production to a halt. Between 2003 and 2007 at Danisco, a Danish yogurt company, researchers in the corporate laboratory were studying this viral infection process and seeking a method to protect the lactic acid–producing bacteria. They discovered a previously unknown mechanism that these bacteria have evolved to fight viral infection. As it happened, these Danisco researchers found a bacterial immunity system, and the repercussions of their discovery have been monumental.

When the Danisco scientists investigated the DNA sequences in lactic acid bacteria that were resistant to phage infection, they found something odd. The bacterial genome contained short repeated DNA sequences that were exact copies of pieces of phage DNA. The bacteria had captured and stored a short fragment of the DNA of past viral invaders in their chromosomes, and they had also evolved a molecular mechanism that used these stored fragments to recognize DNA from new infections and destroy the invading virus. These "immunized" bacteria could quickly produce an RNA copy of the stored viral DNA segment that could find and match the corresponding segment on the foreign viral DNA. The bacteria also produce an enzyme, Cas9, that cuts and destroys the viral DNA at the tagged site. This defense mechanism protects the lactic acid bacteria and serves to immunize the bacterial culture against infection by the viral strains common in their environment. Discovery of this bacterial immunization mechanism, and characterization of its mechanism for editing foreign DNA, soon led other scientists to develop the technology that we now know as *gene editing*. The technique, referred to as CRISPR/Cas9, can modify any target gene in humans, plants, livestock, pathogens, or pathogen vectors. If, for example, a gene contains a mutation that causes an inherited disease, gene editing could destroy this deleterious gene and replace it with a "normal" gene, thereby restoring function.

"Genetic engineering" has been practiced since the eighteenth century, when farmers discovered how to use breeding programs to produce livestock with desired characteristics. But now, our ability to sequence full genomes and to use gene editing to target individual genes opens an entirely new era in the manipulation of the instructions of life. This breakthrough technology expands the tool kit for basic research in living systems. Importantly, gene editing enables the correction of mutations in specific genes that endanger survival of our sources of food and of us as a species.

Gene editing can reengineer the genomes of animal and insect disease vectors that normally harbor the pathogen. In some instances, specific changes in the genome can be propagated rapidly through a population of, for example, the mosquito that carries the Zika virus. Another potential strategy might be to edit the genomes of food plants to generate resistance to pathogens. However, any genetic change made using the CRISPR/Cas9 technology that is propagated to all descendants of the modified plants must be approached with great caution since CRISPR/Cas9 might also introduce changes into the genome far from the targeted site with totally unpredictable consequences to descendant plant generations.

Particularly problematic are the consequences of editing genes in the germ line of humans with the chance of unpredictable off-target changes to the person's DNA. Changes in the chromosomes of eggs, sperm, and embryos would be inherited by all future generations. Before this technology can be widely used to modify heritable characteristics, its level of accuracy must be thoroughly understood. Unintended changes to an individual's genome in somatic cells could be deleterious to that individual, but this would be balanced by saving the life of the individual. However, if unintended edits to DNA in germ line cells and embryos were to be passed down to future generations, these descendants would inherit both the desired beneficial change and an unknown number of unintended and unexpected changes with unpredictable, but possibly harmful, consequences. In response to these concerns, global scientific communities and ethicists are discussing how and whether these gene-editing applications should be controlled. The NIH has called for a moratorium on using NIH funds for editing human embryos. Recently, DARPA (US Defense Advanced Research Projects Agency) has earmarked funds to improve the accuracy of gene editing. The goal is to ensure beneficial use and contain accidental or nefarious misuse.

I'm sitting here thinking this is like the first chapter of Genesis— where the world is created in seven days, and then Adam and Eve say, "We'll take it from here. We thank you for everything up to now." But now with 3D printing and AI and robotics and nukes and gene editing—it's kind of like we're writing the new Bible.   —Bishop William Swing

---

Where Do We Go from Here?

Whether we're dealing with manipulation (and possible eradication) of an entire species of mosquito, generating a prototype pandemic flu strain, or correcting a single mutant gene in a child with cystic fibrosis, any policy decisions will affect all nations and the global population. The breakthroughs in biotechnology and their targets of intervention have global impact, as do any attempts to enact moratoriums on scientific exploration that is believed by some to be dangerous.

It is commonly said that we live now in a global village. The effects of climate change on the geographic distribution of insect and animal vectors and their accompanying viral and bacterial pathogens know no borders. The spread of antibiotic resistance is an irreversible global phenomenon. Unsafe poultry or swine farming practices in, say, Vietnam, China, Greece, or Italy could facilitate the growth of antibiotic-resistant bacteria, and the plasmids carrying the resistance mechanisms can show up in drug-resistant bacteria infecting a child in Toronto. Or an immune-compromised TB patient in, say, an Illinois prison could, just by chance, provide the environment for evolution of a new drug-resistant *Mycobacterium tuberculosis* strain. The corollary to these observations is that the response also must involve an international global effort.

We see some of this happening already. We now have global networks that report outbreaks of disease integrating data provided by the CDC, WHO, the Pasteur Institute International Network, and multiple clinic sites, including those in East Asia, Africa, and South America. During the SARS pandemic there was international cooperation in establishing border monitoring to detect infected persons and follow up with actions to stop spread of the infection. Computer networks and databases are being developed that have the potential to quickly identify and track the spread of new pathogens worldwide. But there is a problem. We can identify new pathogens and we can rather quickly become aware of disease outbreaks anywhere in the world. But that does not mean we have effective means to respond.

Medications to respond to new diseases or to old diseases that have become resistant to established treatments will only come from expensive R&D programs. As we noted, "market solutions" for these problems will not be forthcoming as the business risk and return numbers are not favorable. So, who will provide the necessary funds for disease treatment and containment? Currently, the United States pays 60 percent, the Gates Foundation 10 percent, Britain 13 percent, and five other countries provide the rest of the $4 billion used per year globally for pandemic mitigation. Six nations—the United States, Finland, Saudi Arabia, Pakistan, Eritrea, and Tanzania—have begun establishing fiscal plans for dealing with possible human and animal disease outbreaks. The World Bank supports the creation of emergency funds to deal with disease outbreaks anywhere in the world.

And where will we find the trained and talented research scientists who can develop the solutions to these problems? They will have to come from the international community. In 2012, on international standardized tests, US fifteen-year-olds ranked twenty-first in science and twenty-sixth in math among the thirty-four nations in the Organisation for Economic Co-operation and Development. Over the past twenty years, we have moved from first to tenth place in R&D investment as a percentage of GDP among industrialized nations.[2] There is a great need in these times for farsighted leaders and legislators who can respond with vision and national resources to these looming global health challenges.

In the United States, we need to think much more carefully about what our strategy is for ourselves and the world, and how are we going to take care of ourselves best. There are all kinds of dimensions to that. This topic is one example why acting as though the world doesn't exist is not an option.   —George P. Shultz

---

REFLECTIONS ON DISRUPTION: ECONOMIC GOVERNANCE

John B. Taylor

I am grateful for the opportunity to discuss economic issues in this volume in honor of Sid Drell. Sid's office at the Hoover Institution was just down the hall from mine, and I recall many discussions with him about government policy and economics.

One discussion I recall particularly well. Sid had just read about the so-called Taylor Rule, an equation designed to help the Federal Reserve and other central banks set the interest rate. He found this new "technology" of central banking interesting and asked if I could stop by his office to talk about it, which I did.

I began by writing the equation on the board in his office: 1.5 times one variable, plus 0.5 times another, plus 1. Sid, a physicist who knew much more than me about physical laws and equations, said it couldn't be: "You just can't have a '1' like that. That's not how the world works, John. Maybe an e or a π, or the speed of light as mathematical constants to work with, but not just a '1.'" So, we had a long discussion about where the "1" came from. He was right, of course. It is not a constant of nature, and as it turns out, central banks have been tweaking it recently, trying to improve the equation.

It is, of course, important to examine technological change, as many people here have been doing, and to list the potential problems and to develop remedies to deal with the problems. Indeed, I am struck by the enormous dangers and the worrisome global threats to peace and prosperity from technological changes that have been highlighted here. I am also impressed with the creative thinking about how to look for solutions and what the role of government should be in finding ways to counter these dangers.

Benefits as Well as Costs

But in these remarks, I want to focus more on the benefits of technological change. We should be careful, when we look at "solutions" or even "tweaks" aimed at reducing the dangers from various technological changes, that we don't accidentally kill or stymie something that is extraordinarily beneficial.

Technological advances are beneficial because they can improve people's lives. As an economist I would emphasize that such changes make people more productive. Productivity, by definition, measures the amount of goods or services that can be produced per worker-hour. Technological change increases productivity, and when productivity increases, people are better off—their incomes rise, prices fall, and more goods and services are available. That's the fundamental nature of progress, which in the past fifteen years has been responsible for bringing more than one billion people out of poverty around the world, according to the World Bank. And we hope there are many more to come.

The progress is largely due to increased productivity and the spread of technology. People in both rural and urban areas—farmers, truck drivers, laborers, small producers, shop owners—are using new tools such as cell phones, smartphones, and the internet, and they are learning innovative ways to apply these techniques. It's an incredibly powerful way to reduce poverty and make people's lives better. So, while it's very important to point out the dangers and threats and to find ways to cut them off, I'm concerned that if we implement the wrong policies, we may end up losing the very things that power productivity growth in the future.

Policies make a difference. Looking around the world, you can see good economic performance in some countries and bad economic performance in other countries, and there are changes in countries over time. Many of these differences in performance are due to differences in national economic policies; policies that encourage markets and offer predictable incentives help boost productivity. We should communicate more about what works well. It's not hurting us when more countries do better.

In fact, many people around the world are still very poor. They're not going to get out of poverty unless there's more capital flowing around the world, embodying the technologies such as those we have been talking about at this conference. Capital is naturally a coward: it doesn't want to go places where it's dangerous. It wants to go where it's safe, and where the policies are clear, and the rule of law prevents expropriation. The more that we can have policies like that, the more there'll be the 3D printers, automation, and other technologies from which people can benefit.

Productivity growth is now very low in the United States. It's hard to measure, but that is what the data show. Productivity growth was higher a decade ago. It's gone up and down over the years, and it is down now. So, I think there's a grave danger that we mistakenly stomp out the technology that is going to raise productivity and make people's lives better than in the past. If we forget that, we may introduce controls or regulations that go in the wrong direction, and then we're worse off.

One danger, often pointed out, is that advanced technology will eliminate jobs, with driverless trucks being a prime example. However, when people talk about these issues—and this is a problem in all areas of economics—they tend to pay a lot of attention to the short run and forget about the long run. It used to be that this country had over 50 percent of the workforce involved in agriculture. Over the years amazing technologies came, and now it's only 2 percent producing the same or even more food—a huge increase in productivity. Yet the unemployment rate is no higher now than it was then. So where did all the jobs go? They went to other areas—manufacturing, telecommunications, health care—and they grew in number. Over time it's a huge benefit, though in the shorter run there may be painful adjustment costs especially for people whose job is replaced, say by a tractor, and for this reason we need a good safety net.

The same thing is happening now with other technologies, including new forms of artificial intelligence, and it is likely to continue happening going forward. The good news is that the resulting productivity growth will raise incomes; it will thereby reverse the slow growth of recent years in the United States and other advanced countries and continue relatively high growth in many other parts of the world. Sometimes the beneficial effects are broad-based, and sometimes they are narrowly focused on a few sectors. In the latter case, you don't want to stifle something that's improving people's lives just because it's not universal—you may even be able to find ways to offset the cost by broadening the gains.

It's easy to miss the advantages of technology because they become so routine. Take a hearing aid. It's a simple function, and it may not be as sophisticated as some forms of artificial intelligence. But technological advances in hearing aids can completely change people's ability to function. I wouldn't have been able to become a professor without this technology.

A particularly difficult problem is that technological change often brings inherent costs along with the benefits. Consider Charlie Hill's example in his thoughtful presentation later in this volume. He notes that new communication and social media technologies make it possible for all the "crazy uncles in the attics" to get together and cause mass craziness. That's true, and that is a cost that must be considered. But what about the sane uncles and the other people who can now communicate and collaborate more easily with one another? Today I can easily do research and write papers with economists all over the world, such as my collaborator Volker Wieland in Germany; they do not have to be Stanford colleagues. And the same is true for other areas of human endeavor. We shouldn't get too distracted from broader progress by the edge cases, however meaningful they may be. And there may be ways to both keep the benefits and cut the costs.

Looking for Strategies That Keep the Benefits and Reduce the Harm

One important area of governance where technological change can bring both benefit and harm is the global financial system. I now serve on the Eminent Persons Group on Global Financial Governance, which the Group of 20 countries (G20) recently established. It consists mainly of former finance ministers, central bank governors, and other government finance officials. The purpose of the group is to think about how the international monetary and financial system, including the international financial institutions like the World Bank and the International Monetary Fund (IMF), can be reformed. We are supposed to report back to the G20 countries in a year with recommendations and suggestions.

Early on we found that it is important to establish the goals we are after: what are we trying to achieve? That setting such goals is important may sound obvious, but it's a key issue for this book on governance and worth mentioning here. The first goal is strong economic growth that brings people out of poverty and reduces the income distribution gap globally. The second goal is avoiding crises and recessions, so we don't have terrible events such as the global financial crisis of 2007–9. And the third goal is to mitigate the deleterious international impacts of technology such as cyberattacks and global financial contagion—the sorts of issues being discussed here in this book.

Once you have these goals you can think about a strategy for global financial governance that achieves the goals. Advances in technology enable instantaneous capital and information flows and make the problem more difficult. All over the world, people see the impact on stock or bond prices of a change in monetary or fiscal policy in the United States, Japan, or Europe. Thus, there may be a trade-off between the goals: more open capital markets might be good for growth but may raise issues about cyberattacks. And efforts to control interchanges on social media may help prevent attacks but can also limit the benefits of the spread of technology. Any good strategy needs to be aware of the trade-offs and try to deal with them.

When we were colleagues at the Hoover Institution, Defense Secretary James Mattis would often say, "We live in a strategy-free world," with respect to national security and geopolitics. I think he coined the term "strategy-free." In my view, we also live in a strategy-free world with respect to geoeconomic issues.

It's possible to create a less strategy-free world in global economics and deal with the trade-offs between goals. To do so, one first needs to look for situations or areas where there can be a common strategic approach, where each nation is basically thinking about governing its own interests, but where that is also conducive to good governance globally. For example, in the past five or so years there have been many unconventional monetary policy actions taken by the central banks of the United States, Japan, and Europe. These policies affected exchange rates, with the dollar depreciating at first, then the yen, and then the euro. The movements are huge, and they affect people's decisions about foreign investment and about exports and imports. Some argue that the effects on exchange rates are so large that policy-makers at the international financial institutions should limit the flow of capital—sometimes called "capital flow management."

But if each country, if each central bank, said what its own monetary strategy was for its own country, and it was a clear strategy, it would help create global stability and reduce that exchange rate volatility quite a bit. It's not just the quality of the strategy itself that is valuable, but also the very statement of what the strategy is in the first place. Then people operating in financial markets, including policy-makers in other countries, could observe this and say, "Well, that's their strategy. Here's our strategy. Here's our method for adjusting interest rates," or whatever the policy instrument should be. You're trying to just understand what the others are doing. This strategy would limit the harm from excessive exchange rate volatility without choking off beneficial capital flows.

New financial technologies can help. One potentially attractive element of blockchain technology, for example, is that it gives ways to make payments between people or between firms that are independent of government actions. Central banks and governments are examining blockchain technology to see if they can use it themselves. This technology may lead to better policy. We're not there yet, but you could imagine large potential benefits.

People sometimes worry about financial technology upending the financial order, perhaps driving out the US dollar. But the dollar, viewed in a historical context, is amazingly resistant to changes like this—whether it's in competition with the euro, China's RMB, or even bitcoin over time. And I think that will continue, even if the technology changes, if the United States stays with good economic policy.

Government versus Private Remedies

I experienced four different stints in the federal government over the years, so I know it is sometimes hard to get things done. The private sector has advantages in many respects—flexible salaries, the ability to fire people, more incentives. Plus, people in government don't always have the training in the newest technologies. So there may be limits to what the government can do to design the types of strategies I am suggesting here. Often, well-intentioned regulations impede improvement, keeping new products from coming into play. This suggests looking for private remedies where possible.

In addition, there are innovative ways to use technology, like artificial intelligence, to make government regulations more flexible—a more efficient way to get the same result. Of course, it would be dangerous, especially if the stakes are big, to just apply an algorithm to the private sector without considering robustness. You need to bring in judgment from knowledge of history, of international relations, and the nature of human conflict alongside that.

Considering policy responses to technology change, some have argued for a universal basic income to address labor force disruptions from technology. But I do not think we want a society where work isn't valued. People should feel that they are contributing, and they need to participate to be part of the social fabric.

Others have said we should tax the robots as a solution. But a robot is a machine, a tool, a piece of capital. Taxing that is simply taxing something that can improve people's lives. As I've already mentioned, US productivity is already quite low, reflected in our low overall economic growth rate, and it's largely because people are not investing enough in capital, or in the right places, to improve it. If you are going to go down the route of taxing robots to try to slow down the problem, there is less of a chance that people will take on the task of really fixing the core things that need to be fixed.

Education and Technological Change

I also want to touch on education, a topic that various authors have raised. Education could benefit tremendously from technology. But we need to introduce organizational changes in the education system and use incentives, perhaps making better use of the financial resources we already have dedicated toward our educational system. This is an area that could benefit from artificial intelligence and technology more broadly. Many teachers and professors are using the same techniques they have been using in teaching for many years. But technology enables a student in a poor region without access to good schools or colleges to take courses online from the best teachers in the world.

I have an online version of the introductory economics course I teach at Stanford. It's a very rewarding experience when tens of thousands are following your course around the world. An interesting thing I noticed in teaching the course this past summer was that there were people over one hundred years old taking the course online. I had thought that the audience consisted mainly of college students, but the *average* age was over thirty-five.

I inquired about other online courses offered by Stanford, and I found a similar age distribution in these other courses. This is not, of course, a bad thing. It suggests that online courses can be of great benefit without being disruptive. Recognizing that people are going to be moving in and out of the labor force after their formal education, it suggests a way to maximize benefits and minimize the costs of the introduction of new potentially disruptive technologies. I hope that the education system is flexible and adapts to this new mode of learning.

More generally, the key is to make the economy more flexible. You cannot rely solely on command and control. It's hard to know what the technology is going to be ten to twelve years from now, so in terms of skills to teach people to anticipate change, it's hard to just point to STEM or the humanities. People need to learn how to be flexible. And for people to be flexible, policy should be flexible.

The financial technology industry, for example, suffers from the possibility of being overregulated, such that its benefits are snuffed out, and that's a danger. But there's some sympathy from people at traditional financial firms who feel there's already too much regulation in the industry. So, this could be a place where there is some convergence of views between "fin tech" people, say in Silicon Valley, and other more traditional financial people around the country. It is an example of a way to get things done in the face of technical change.

Conclusion

To end on somewhat of a light note, I'll share one simple story about the benefits and harms of technological change and what to do about it. The story is due to Bruno Bettelheim and is told in his book *The Informed Heart: Autonomy in a Mass Age*. It illustrates how you can learn to live with technological change over time and overcome unexpected bumps along the way.

The story goes back to the time when automatic dishwashers were invented. A husband and wife buy a new dishwasher, and suddenly they find their marriage is falling apart. They go to a counselor who asks them if anything at home has changed. "Well," replies the couple, "we bought a new dishwasher." "Did you used to talk to each other while you hand-washed the dishes every night?" "Yeah, we did!" they answer. "And what about now?" "Well, no, now we just put the dishes in and press the button."

So, the counselor suggests to the couple a simple solution. Sit down and make an extra effort to talk while the dishwasher is running. It is an obvious strategy, but they try it, and it works.

The lesson: sometimes it's not easy to anticipate or even understand the costs of a technological advance and how it can change your life. But by putting an emphasis on finding a strategy to adapt we can eliminate, or at least minimize, the harms of the technology while taking advantage of the benefits.

6

GOVERNANCE AND ORDER IN A NETWORKED WORLD

Niall Ferguson

"Is the world slouching toward a grave systemic crisis?" asked the historian Philip Zelikow at the annual gathering of the Aspen Strategy Group earlier this summer, the kind of "deep system-wide crisis . . . when people all over the world no longer think the old order work[s]."[1] Among the reasons he gave for anticipating such a crisis was "the digital revolution and the rise of a networked world." To grasp the scale and nature of this coming crisis, we must begin by recognizing how drastically the balance of power has shifted in our time from hierarchically ordered empires and superpowers (the euphemism for empire developed to suit American and Soviet sensibilities) to distributed networks.[2]

To be sure, the formal "org. chart" of global power is still dominated by the vertically structured super-polities that gradually evolved out of the republics and monarchies of early modern Europe, the colonies they established in the New World, and the older empires of Asia. Though not the most populous nation in the world, the United States is certainly the world's most powerful state, despite—or perhaps because of—the peculiarities of its political system. Its nearest rival, the People's Republic of China, is usually seen as a profoundly different kind of state, for while the United States has two major parties, the People's Republic has one, and only one. The US government is founded on the separation

of powers, not least the independence of its judiciary; the PRC subordi-
nates all other institutions, including the courts, to the dictates of the
Communist Party. Yet both states are republics, with roughly comparable vertical structures of administration and not wholly dissimilar con-
centrations of power in the hands of the central government relative to
state and local authorities. Economically, the two systems are certainly
converging, with China looking ever more to market mechanisms,
while the US federal government in recent years has steadily increased
the statutory and regulatory power of public agencies over producers
and consumers. And, to an extent that disturbs libertarians on both left
and right, the US government exerts control and practices surveillance
over its citizens in ways that are functionally closer to contemporary
China than to the America of the Founding Fathers. In these respects,
"Chimerica" is no chimera. Once these economies seemed like opposites, with one doing the exporting, the other the importing, one doing
the saving, the other the consuming.[3] Since the financial crisis, however,
there has been a certain convergence. Today the real estate bubble, the
excessive leverage, the shadow banks—not to mention the technology
"unicorns"—are almost as likely to be encountered in China as in America. In Chimerica 1.0, opposites attracted. In Chimerica 2.0, the odd
couple have become strangely alike, as often happens in a marriage.

Sitting alongside the United States and the People's Republic in the
hierarchy of nation-states are the French Republic, the Russian Federation, and the United Kingdom of Great Britain and Northern Ireland.
These are the five permanent members of the United Nations Security
Council, and they are thereby set above all the other 188 members of the
UN—an institution in which all nations are equal, but some are more
equal than others. However, that is clearly not a sufficient description of
today's world order. In terms of military capability, there is another,
somewhat larger elite of nuclear powers to which, in addition to the
"P5," also belong India, Israel, Pakistan, and North Korea. Iran aspires
to join them. In terms of economic power, the hierarchy is different
again. The Group of Seven countries (Canada, France, Germany, Italy,
Japan, the United Kingdom, and the United States) were once consid-

cred the dominant economies in the world, but today that club is relatively less dominant as a result of the rise of the "BRICS" (Brazil, Russia, India, China, and South Africa), the biggest of the so-called emerging markets. The Group of 20 was formed in 1999 to bring most of the world's big economies together, but with the Europeans overrepresented (as the EU is a member in its own right, as are the four biggest EU member-states).

Yet to think of the world only in such terms is to overlook its profound transformation by the proliferation of distributed networks in the past forty years. Picture, instead, a network graph (similar to that depicted by figure 6.1) based on economic complexity and interdependence that delineates the relative sophistication of all the world's economies in terms of technological advancement as well as their connectedness through trade and cross-border investment. Such a graph would have a strongly hierarchical architecture because of the power-law-like distribution of economic resources and capabilities in the world and the significant variation in economic openness between countries. Yet it would also unmistakably be a network, with most nodes connected to the rest of the world by more than one or two edges.[4]

Even more striking is the rise of an entirely new global economy based on the internet and composed of "bits" as opposed to "atoms." Amazon began as an online bookstore in Seattle in 1995. Today it has more than three hundred million users and the largest revenues of any internet company in the world. Google started life in a garage in Menlo Park, California, in 1998. Today it has the capacity to process over 4.2 billion search requests every day. In 2005 YouTube was a start-up in a room above a pizzeria in San Mateo, California. Today it allows people to watch 8.8 billion videos a day. Facebook was dreamed up at Harvard just over a decade ago. Today it has close to two billion users who log on at least once a month—more than the population of China.[5] In the United States, Facebook penetration is as high as 82 percent of internet users between the ages of eighteen and twenty-nine, 79 percent of those age thirty to forty-nine, 64 percent of the fifty to sixty-four age group, and 48 percent of those sixty-five and older. If there are six degrees of

FIGURE 6.1   Economic Complexity

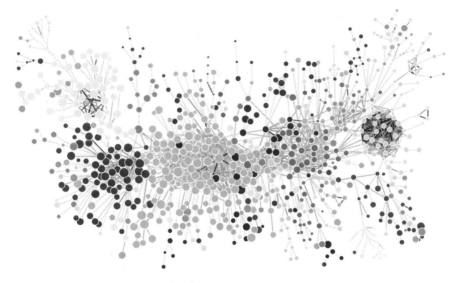

Graph of the global export "product space," where dot size is proportional to total world trade in that good. Dots are shaded according to product type. The central component is dominated by "machinery and electrical" and "transportation" (including cars); the right-hand cluster is the textile, footwear, and headgear industry.

Source: Alexander J. G. Simoes and Cesar A. Hidalgo, "The Economic Complexity Observatory: An Analytical Tool for Understanding the Dynamics of Economic Development," 2011, workshops at the twenty-fifth AAAI Conference on Artificial Intelligence, https://pdfs .semanticscholar.org/7733/68ce1faa36d9ac833b3c3412d136033b91c1.pdf.

separation for humanity as a whole, for Facebook users the average figure is now 3.57.

The key question is how far this networking of the world now poses a threat to the hierarchical world order of nation-states comparable to the threat that online social networks have recently posed to established domestic-political hierarchies—notably in 2011 in the Middle East, in 2014 in Ukraine, in 2015 in Brazil, and in 2016 in Britain and America. To put the question more simply: can a networked world have order? Some—notably Anne-Marie Slaughter—say that it can.[6] In the light of historical experience, I very much doubt it.

> *It seems to me that in this information age, people know what's going on everywhere pretty fast, and they can communicate. Everybody has a cell phone. They can organize—and they do. This means that since there's diversity everywhere, you can't suppress it, or ignore it, or if you try to, it breaks you. So you have to learn how to govern over diversity.*   —George P. Shultz

* * *

According to folklore, Mahatma Gandhi was once asked by a reporter what he thought of Western civilization. He replied that he thought it would be a good idea. The same might be said about world order. In his book of that title, Henry Kissinger argues that the world is in a parlous condition verging on international anarchy. Four competing visions of world order—the European, the Islamic, the Chinese, and the American—are each in varying stages of metamorphosis, if not decay. Consequently, there is no real legitimacy to any of these visions. The emergent properties of this new world disorder are the formation of regional blocs and the danger that friction between them might escalate into some kind of large-scale conflict, comparable in its origins and potential destructiveness to the First World War.[7] Contrary to those who claim (on the basis of a misreading of statistics of conflict) that the world is steadily becoming more peaceful and that "wars between states . . . are all but obsolete," Kissinger argues that the contemporary global constellation of forces is in fact highly flammable.[8] First, whereas "the international economic system has become global . . . the political structure of the world has remained based on the nation-state." (This was a tension laid bare in the 2008 financial crisis when, as governor of the Bank of England, Mervyn King wittily remarked that international banks were "global in life, but national in death.") Second, we are acquiescing in the proliferation of nuclear weapons far beyond the Cold War "club," thus "multiply[ing] the possibilities of nuclear confrontation." Finally, we now have the new realm of cyberspace, which Kissinger likens to Hobbes's "state of nature" in which "asymmetry and a kind of congenital world disorder are built into relations between . . . powers."[9]

Kissinger's warning cannot be lightly dismissed. The world today frequently resembles a giant network on the verge of a cataclysmic outage. Globalization is in crisis. Populism is on the march. Authoritarian states are ascendant. Technology meanwhile marches inexorably ahead, threatening to render most human beings redundant or immortal or both. How do we make sense of all this? In pursuit of answers, many commentators resort to crude historical analogies. To some, Donald Trump is Hitler, about to proclaim an American dictatorship.[10] To others, Trump is Nixon, on the verge of being impeached.[11] But it is neither 1933 nor 1973 all over again. Easily centralized technology made totalitarian government possible in the 1930s. Forty years later, it had already become much harder for a democratically elected president to violate the law with impunity. Nevertheless, the media in the 1970s still consisted of a few television networks, newspapers, and press agencies. And in more than half the world those organs were centrally controlled. It is impossible to comprehend the world today without understanding how it has changed as a result of new information technology. This has become a truism. The crucial question is: *How* has it changed? The answer is that technology has enormously empowered distributed networks of all kinds relative to traditional hierarchical power structures—but that the consequences of that change will be determined by the structures, emergent properties, and interactions of these networks.

The global impact of the internet has few analogues in history better than the impact of printing on sixteenth-century Europe. The personal computer and smartphone have empowered the individual as much as the pamphlet and the book did in Luther's time. Indeed, the trajectories for the production and price of PCs in the United States between 1977 and 2004 are remarkably similar to the trajectories for the production and price of printed books in England from 1490 to 1630 (see figure 6.2).[12] In the era of the Reformation and thereafter, connectivity was enhanced exponentially by rising literacy, so that a growing share of the population was able to access printed literature of all kinds, rather than having to rely on orators and preachers to hear new ideas.

FIGURE 6.2   Prices and Quantities of Books and PCs, 1490s–1630s and 1977–2004

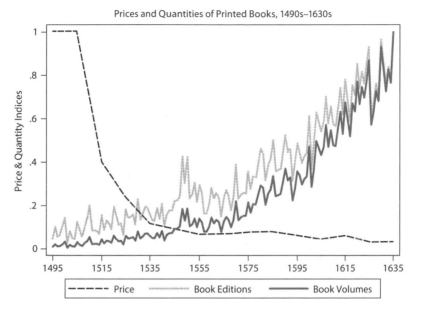

Prices and Quantities of Printed Books, 1490s–1630s

Prices and Quantities of Personal Computers, 1977–2004

Source: Jeremiah Dittmar, February 2012, "The Welfare Impact of a New Good: The Printed Book," working paper.

There are three major differences between our networked age and the era that followed the advent of European printing. First, and most obviously, our networking revolution is much faster and more geographically extensive than the wave of revolutions unleashed by the German printing press. In a far shorter space of time than it took for 84 percent of the world's adults to become literate, a remarkably large proportion of humanity has gained access to the internet. As recently as 1998, only around 2 percent of the world's population was online. Today the proportion is two in five. The pace of change is roughly an order of magnitude faster than in the post-Gutenberg period: what took centuries after 1490 took just decades after 1990. The rate of growth of the global network may be slowing in terms of the numbers of new internet users and smartphone owners added each year, but it shows no sign of stopping. In other respects—for example, the transitions from text to image and video and from keyboard to microphone interface—it is speeding up. Literacy will ultimately cease to be a prerequisite for connectedness.

Nor is this technological revolution confined to developed countries. In terms of connectivity, if little else, the world's poor are catching up fast. Among the poorest 20 percent of households in the world, roughly seven out of ten have cell phones. The Indian telecom company Bharti Airtel has a customer base as large as the US population. Indeed, the number of internet users in India now exceeds that in America. It took just eight years for all Kenyan households (and close to 90 percent of individuals) to have cell phones. It took four years for Safaricom's pioneering M-Pesa payment system to reach 80 percent of households.[13] Even impoverished and chaotic Somalia went from 5 to 50 percent cell phone penetration inside five years.[14] Selling the world's poor mobile telephony is proving easier than providing them with clean water—a strong argument for leaving the provision of clean water to the private sector rather than weak, corrupt governments.[15]

Second, the distributional consequences of our revolution are quite different from those of the early modern revolution. Late-fifteenth-century Europe was not an ideal place to enforce intellectual property rights, which in those days existed only when technologies could be secretively monopolized by a guild. The printing press created no billionaires: Gutenberg was not Gates. Moreover, only a subset of the media made possible by the printing press—newspapers and magazines—sought to make money from advertising, whereas the most important ones made possible by the internet do. Nevertheless, few people foresaw that the giant networks made possible by the internet, despite their propaganda about the democratization of knowledge, would be profoundly inegalitarian. A generation mostly removed from conflict—the baby boomers—had failed to learn the lesson that it is not unregulated networks that reduce inequality but wars, revolutions, hyperinflations, and other forms of expropriation.[16]

To be sure, innovation has driven down the costs of information technology. Globally, the costs of computing and digital storage fell at annual rates of 33 and 38 percent per annum between 1992 and 2012.[17] However, contrary to the hopes of those who envisioned a big bazaar of crowd-sourced applications, the internet has evolved into a vast scale-free network, complete with hyperconnected nodes that are more like super-hubs.[18] Oligopolies have developed in the realms of both hardware and software, as well as service provision and wireless networks. The nexus between the seemingly indestructible AT&T and the reinvented Apple illustrates an old truth: corporations will pursue monopoly, duopoly, or oligopoly if they are left free to do so. Even those corporations committed to an "open architecture" web—such as Amazon, Facebook, and Google—seek monopolistic power in their segments: respectively, e-commerce, social networks, and search.[19] Poor governance and regulation explain huge differentials in cellular service and internet costs between countries.[20] They also explain why a small number of countries dominate the information and communications technology industry (though it is striking that the United States ranks seventh—some way behind Ireland, South Korea, Japan, and the UK—in terms of the relative importance of information technology to its economy as a whole).[21]

These dynamics explain why the ownership of the world's electronic network is so concentrated. As of the fall of 2017, Google (or rather the renamed parent company Alphabet Inc.) is worth $665 billion by market capitalization. Around 11 percent of its shares, worth around $76 billion, are owned by its founders, Larry Page and Sergey Brin. The market capitalization of Facebook is approaching $500 billion; 14 percent of the shares, worth $71 billion, are owned by its founder, Mark Zuckerberg. Despite their appearance as great levelers, social networks are thus, in the words of social network theoretician Charles Kadushin, "inherently unfair and exclusionary."[22] Because of preferential attachment—the tendency for well-connected hubs to get even better connected—the ultimate social network truism does indeed come from the book of Matthew: "For unto every one that hath shall be given, and he shall have abundance: but from him that hath not shall be taken away even that which he hath" (Matthew 25:28–29). Unlike in the past, there are now two kinds of people in the world: those who own and run the networks and those who merely use them. The commercial masters of cyberspace may still pay lip service to a flat world of netizens, but in practice, companies such as Google are hierarchically organized, even if their org. charts are quite different from those of General Motors in Alfred Sloan's day.

In traditional societies, the advent of market forces disrupts often hereditary networks and as a result promotes social mobility and reduces inequality. Meritocracy prevails. But when networks and markets are aligned, as in our time, inequality explodes as the returns on the network flow overwhelmingly to the insiders who own it.[23] Granted, the young and very wealthy people who own the modern networks tend to have somewhat left-wing political views. (Peter Thiel is the rare exception: a libertarian who was willing to sup with the populists in 2016.) However, few of them would welcome Scandinavian rates of personal income tax, much less an egalitarian revolution. The masters of the internet would seem to relish being rich almost as much as the wolves of precrisis Wall Street a decade ago, though their consumption is less conspicuous than their pangs of conscience. It is hard to imagine an investment banker following the example of Sam Altman of Y Combinator and going on a pilgrimage to Middle America, as if doing penance for the 2016 election result.[24] Yet the San Francisco to which Altman returns remains a city of glaring inequality, not least because of the distortions that ensure that decent housing is ludicrously expensive. (Ownership of real property is second only to ownership of intellectual property as a determinant of wealth inequality, but the most valuable housing is located closest to the geographical clusters where the most valuable intellectual property is generated.) And all that the big technology companies seem willing to offer the millions of truck and taxi drivers they intend to replace with driverless cars is some form of basic income. The sole consolation is that the largest shareholders of the FANG companies (Facebook, Amazon, Netflix, and Google) are US institutional investors, which, insofar as they are the managers of the savings of the American middle class, have therefore given that class a significant stake in the profits of the information technology industry. An important qualification, however, is that foreign investors probably own at least 14 percent of the equity of major US corporations and, in the case of companies with very large foreign sales (such as Apple, which earns around two-thirds of its revenue abroad), almost certainly much more.[25] No serious student of capital markets, however, would attribute to these foreign investors even a shred of influence over the companies' corporate governance.

Thirdly, and finally, the printing press had the effect of disrupting religious life in western Christendom before it disrupted anything else. By contrast, the internet began by disrupting commerce; only very recently did it begin to disrupt politics, and it has only really disrupted one religion, namely Islam.

Generations succeeding generations, technology and the humanities grow farther apart. Technology took the lead, and technology has now become the master of the scene.   —Charles Hill

---

* * *

In many ways, networks were the key to what happened in American politics in 2016. There was the grassroots network of support that Donald Trump's campaign built—and that built itself—on the platforms of

Facebook, Twitter, and Breitbart. These were, to use Trump's words, the "forgotten" men and women who turned out on November 8 to defeat the "global special interests" and "failed and corrupt political establishment" that Trump's opponent Hillary Clinton was alleged to personify. A role was also played by the jihadist network, as the Islamic State–affiliated terror attacks during the election year lent credibility to Trump's pledges to "strip out the support networks for radical Islam in this country" and to ban Muslim immigration.

As a very wealthy man who could nevertheless play the role of demagogue with aplomb, Trump himself embodied a central paradox of the age. He was at once a minor oligarch and a major brand. Trump is perhaps unique in having assumed the office of president with a tangled network of businesses, investments, and corporate links—to as many as 1,500 people and organizations.[26] At the same time, Trump's campaign succeeded where his opponents failed in harnessing the networks of Silicon Valley, to the dismay of the people who owned and thought they also controlled the networks. Their agony in the weeks after the election was palpable. Google at first sought to woo the new administration, only to denounce its executive orders limiting travel and migration to the United States from certain Muslim-majority countries.[27] Mark Zuckerberg absented himself from a meeting with the new president attended by other technology CEOs. Presumably it was some comfort to him that the Women's March against Trump had also organized itself through Facebook.[28] It is hard to believe that there will not be some kind of clash between the Trump administration and the big information and communications technology companies, especially if the administration overturns its predecessor's decision in 2015 that the Federal Communications Commission should regulate the internet as a public utility, like the old railroad or telephone networks. There seems an obvious conflict of interest between telecom and cable companies and bandwidth-greedy platform content providers such as Netflix over the issue of "net neutrality" (the principle that all bits of data should be treated alike, regardless of their content or value).[29] Antitrust action against the FANG companies could be Trump's next move, though it would be surprising if a Republican administration went down that route. What seems highly unlikely is that Silicon Valley companies will continue to enjoy the exemptions from being treated as publishers that they were granted under the Telecommunications Act (1996).

Yet in two respects there is a clear similarity between our time and the revolutionary period that followed the advent of printing. Like the printing press, modern information technology is transforming not only the market—most recently, by facilitating the sharing (i.e., short-term rentals) of cars and apartments—but also the public sphere. Never before have so many people been connected together in an instantly responsive network through which "memes" can spread even more rapidly than natural viruses.[30] However, the notion that taking the whole world online would create a utopia of netizens, all equal in cyber-space, was always a fantasy—as much a fantasy as the Lutheran vision of a priesthood of all believers. The reality is that the global network has become a transmission mechanism for all kinds of manias and panics, just as the combination of printing and literacy for a time increased the prevalence of millenarian cults and witch crazes. The cruelties of the Islamic State seem less idiosyncratic when compared with those of some governments and sects in the sixteenth and seventeenth centuries.[31]

Secondly, as in the period during and after the Reformation, our time is seeing an erosion of territorial sovereignty.[32] In the sixteenth and sev-enteenth centuries, Europe was plunged into a series of religious wars because the principle formulated at the Peace of Augsburg (1555)—*cuius regio, eius religio*, "to each ruler, the religion he chooses"—was honored mainly in the breach. In the twenty-first century, we see a similar phenomenon of escalating intervention in the domestic affairs of sovereign states.

There was, after all, a third network involved in the US election of 2016, and that was Russia's intelligence network. At the time of writing, it is clear that the Russian government did its utmost to maximize the damage to Hillary Clinton's reputation stemming from her and her cam-paign's sloppy email security, using WikiLeaks as the conduit through which stolen documents were passed to the American media. To visit the WikiLeaks website is to enter the trophy room of this operation. Here is the Hillary Clinton Email Archive, there are the Podesta Emails. Not all the leaked documents are American, to be sure. But you will look in vain for leaks calculated to embarrass the Russian government. Julian Assange may still skulk in the Ecuadorean embassy in London, but the reality is that he lives, an honored guest of President Vladimir Putin, in the strange land of Cyberia—the twilight zone inhabited by Russia's online operatives.

Russian hackers and trolls pose a threat to American democracy similar to the one that Jesuit priests posed to the English Reformation: a threat from within sponsored from without. "We're at a tipping point," according to Admiral Michael S. Rogers, head of the National Security Agency and US Cyber Command.[33] Cyberactivities are now at the top of the director of national intelligence's list of threats. And WikiLeaks is only a small part of the challenge. The Pentagon alone reports more than ten million attempts at intrusion each day.[34] Of course, most of what the media call cyberattacks are merely attempts at espionage. To grasp the full potential of cyberwarfare, one must imagine an attack that could shut down a substantial part of the US power grid. Such a scenario is not far-fetched. Something similar was done in December 2015 to the Ukrainian electricity system, which was infected by a form of computer malware called BlackEnergy.

Computer scientists have understood the disruptive potential of cyberwarfare since the earliest days of the internet. At first it was adolescent hackers who caused mayhem, geeks like Robert Tappan Morris, who almost crashed the World Wide Web in November 1988 by releasing a highly infectious software worm.[35] Another was "Mafia Boy," the fifteen-year-old Canadian who shut down the Yahoo website in February 2000. Blaster, Brain, Melissa, Iloveyou, Slammer, Sobig—the names of the early viruses betrayed their authors' youth.[36] It is still the case that many cyberattacks are carried out by nonstate actors: teenage vandals, criminals, "hacktivists," or terrorist organizations. (The October 21, 2016, attack launched against the domain name service provider Dynamic Network Services Inc., which used Chinese-manufactured webcams as "bots," was almost certainly a case of vandalism.[37]) However, the most striking development of 2016 was the rise of Cyberia.

As the country that built the internet, the United States was bound to lead in cyberwarfare, too. It began to do so as early as the first Reagan administration.[38] During the 2003 Iraq invasion, US spies penetrated Iraqi networks and sent messages urging generals to surrender.[39] Seven years later it was the United States and Israel that unleashed the Stuxnet virus against Iran's nuclear enrichment facilities.[40] The problem is not just that two can play at that game. The problem is that no one knows how many people can play at any number of cybergames. In recent years, the United States has found itself under cyberattack from Iran, North Korea, and China. However, these attacks were directed against companies (notably Sony Pictures), not the US government. The Russians were the first to wage war directly against the US government, seeking to compensate for their relative economic and military decline by exploiting the "wide asymmetrical possibilities" that the internet offers for "reducing the fighting potential of the enemy."[41] They learned the ropes in attacks on Estonia, Georgia, and Ukraine. In 2016, however, the Kremlin launched a sustained assault on the American political system, using as proxies not only WikiLeaks but also the Romanian blogger "Guccifer 2.0."[42]

Let us leave aside the question of whether or not the Russian interference—as opposed to the fake news discussed in the previous chapter—decided the election in favor of Trump; suffice to say it helped him, though both fake and real news damaging to Clinton could presumably have been disseminated without Russia's involvement. Let us also leave aside the as yet unresolved questions of how many members of the Trump campaign were complicit in the Russian operation, and how much they knew.[43] The critical point is that Moscow was undeterred. For specialists in national security, this is only one of many features of cyberwar that are perplexing. Accustomed to the elegant theories of "mutually assured destruction" that evolved during the Cold War, they are struggling to develop a doctrine for an entirely different form of conflict in which there are countless potential attackers, many of them hard to identify, and multiple gradations of destructiveness. As then deputy secretary of defense William Lynn observed in 2010, "Whereas a missile comes with a return address, a computer virus generally does not." For Joseph Nye of Harvard's Kennedy School, deterrence may be salvageable, but that is true only if the United States is prepared to make an example of an aggressor. The three other options Nye proposes are to ramp up cybersecurity, to try to "entangle" potential aggressors in trade and other relationships (so as to raise the cost of cyberattacks to them), and to establish global taboos against cyber akin to the ones that have (mostly) discouraged the use of biological and chemical weapons.[44] This analysis is not very comforting. Given the sheer number of cyber-aggressors, defense seems doomed to lag behind offense, in an inversion of conventional military logic. And the Russians have proved themselves to be indifferent to both entanglement and taboos, even if China may be more amenable to Nye's approach. Indeed, the Russian government seems willing to enter into partnerships with organized criminals in pursuit of its objectives.[45]

How frightened should we be of Cyberia? For Anne-Marie Slaughter, our hypernetworked world is, on balance, a benign place, and the "United States . . . will gradually find the golden mean of network power."[46] True, there are all kinds of networked threats ("terrorism . . . drug, arms, and human trafficking . . . climate change and declining biodiversity . . . water wars and food insecurity . . . corruption, money laundering, and tax evasion . . . pandemic disease"). But if America's leaders can only "think in terms of translating chessboard alliances into hubs of connectedness and capability," all should come right. The key, she argues, is to convert hierarchies into networks, turning NATO into "the hub of a network of security partnerships and a center for consultation on international security issues," and reforming the United Nations Security Council, the International Monetary Fund, and the World Bank by opening them up to "newer actors."[47] The institutions of world order established after the Second World War need to metamorphose into "hubs of a flatter, faster, more flexible system, one that operates at the level of citizens as well as states," incorporating "good web actors, corporate, civic, and public." One example she gives is the Global Covenant of Mayors for Climate and Energy, which connects more than 7,100 cities around the world.[48] Another is the Open Government Partnership launched by the Obama administration in 2011, which now includes seventy countries committed to "transparency, civic participation, and accountability."[49]

Can the "good actors" join together in a new kind of geopolitical network, pitting their "webcraft" against the bad actors? Joshua Cooper Ramo is doubtful. He agrees with Slaughter that "the fundamental threat to American interests isn't China or al-Qaeda or Iran. It is the evolution of the network itself." However, he is less sanguine about how easily the threat can be combated. Cyberdefense lags ten years behind cyberattack, not least because of a new version of the impossible trinity: "Systems can be fast, open, or secure, but only two of these three at a time."[50] The threat to world order can be summed up as "very fast networks x artificial intelligence x black boxes x the New Caste x compression of time x everyday objects x weapons."[51] In *The Seventh Sense*, Ramo argues for the erection of real and virtual "gates" to shut out the Russians, the online criminals, the teenage net vandals, and other malefactors. Yet he himself quotes the three rules of computer security devised by the NSA cryptographer Robert Morris, Sr.: "RULE ONE: Do not own a computer. RULE TWO: Do not power it on. RULE THREE: Do not use it."[52] If we all continue to ignore those new categorical imperatives—and especially our leaders, most of whom have not even enabled two-factor authentication on their email accounts—how will any gates keep out the likes of Assange and Guccifer?

An intellectual arms race is now under way to devise a viable doctrine of cybersecurity. It seems unlikely that those steeped in the traditional thinking of national security will win it. Perhaps the realistic goal is not to deter attacks or retaliate against them but to regulate all the various networks on which our society depends so that they are resilient—or, better still, "anti-fragile," a term coined by Nassim Taleb to describe a system that grows stronger under attack.[53] Those like Taleb who inhabit the world of financial risk management saw in 2008 just how fragile the international financial network was: the failure of a single investment bank nearly brought the whole system of global credit tumbling down. The rest of us have now caught up with the bankers and traders; we are all now as interconnected as they were a decade ago. Like the financial network, our social, commercial, and infrastructural networks are under constant attack from fools and knaves, and there is very little indeed that we can do to deter them. The best we can do is to design and build our networks so that they can withstand the ravages of Cyberia. That means resisting the temptation to build complexity when (as in the case of financial regulation) simplicity is a better option.[54] Above all, it means understanding the structures of the networks we create.

When half the nodes of a random graph the size of most real-world networks are removed, the network is destroyed. But when the same procedure is carried out against a scale-free model of a similar size, "the giant connected component resists even after removing more than 80 per cent of the nodes, and the average distance within it [between nodes] is practically the same as at the beginning."[55] That is a vitally important insight for those whose task is to design networks that can be anti-fragile in the face of a deliberate, targeted attack.

Dictatorships are now empowered by this electronic revolution. They now have technologies that were way beyond their wildest, happiest dreams before, to enable them.   —Charles Hill

---

* * *

In March 2017 the members of the House of Commons Home Affairs Committee, led by its chair Yvette Cooper, attacked Google, Facebook, and Twitter for not doing enough to censor the internet on their behalf. Cooper complained that Facebook had failed to take down a page with the title "Ban Islam." As she put it, "We need you to do more and to have more social responsibility to protect people."[56] In the same week, the

German justice minister Heiko Maas unveiled a draft law that would impose fines of up to fifty million euros on social networks that fail to delete "hate speech" or "fake news." In his words, "Too little illegal content is being deleted and it's not being deleted sufficiently quickly."[57]

One can argue for and against censorship of odious content. One can marvel that companies and government agencies would spend money on online advertising so indiscriminately that their carefully crafted slogans end up on jihadist websites. However, arguing that Google and Facebook should do the censoring is not just an abdication of responsibility; it is evidence of unusual naïveté. As if these two companies were not already mighty enough, European politicians apparently want to give them the power to limit their citizens' free expression.

There are three essential points to understand about the IT revolution. The first is that it was almost entirely a US-based achievement, albeit with contributions from computer scientists who flocked to Silicon Valley from all over the world and Asian manufacturers who drove down the costs of hardware. Second, the most important of the US tech companies are now extraordinarily dominant. Third, as we have seen, this dominance translates into huge amounts of money. Confronted with this American network revolution, the rest of the world had two options: capitulate and regulate or exclude and compete. The Europeans chose the former. You will look in vain for a European search engine, a European online retailer, a European social network. The biggest EU-based internet company is Spotify, the Stockholm-based music and video streaming company founded in 2006.[58] The FANG has been sunk deep into the EU, and all the European Commission can do now is to harass the US giants with antitrust charges, backdated tax bills, and tighter rules on privacy and data protection, not to mention employment rights.[59] To be sure, the Europeans led the way in insisting that American companies could not operate in their territory independently of national or European law. It was a Frenchman, Marc Knobel, who established that Yahoo could not advertise Nazi memorabilia on its auction sites, not least because the server through which French users accessed the site was located in Europe (in Stockholm), but also because

Yahoo was not (as it claimed) incapable of distinguishing French from other users.[60] A number of European countries—not only France but also Britain and Germany—have passed laws that require internet service providers to block proscribed content (such as pedophile pornography) from being viewed by their citizens. Yet the European political elites now effectively rely on US companies such as Facebook to carry out censorship on their behalf, seemingly oblivious to the risk that Facebook's "community standards" may end up being stricter than European law.

The Chinese, by contrast, opted to compete. This was not the response predicted by Americans, who assumed that Beijing would simply try to "control the Internet"—an endeavor President Bill Clinton famously likened to "trying to nail Jell-O to the wall."[61] "The Internet is a porous web," wrote one American academic in 2003, "and if people in China . . . want to get information from sites in Silicon Valley, even the most omnipotent of governments will be hard pressed to stop them."[62] This was not quite right. Certainly, there has been censorship. Since 2012, when Lu Wei was put in charge of the Central Leading Group for Cyberspace Affairs, China has increased the effectiveness of its Great Firewall, which blocks access to tens of thousands of Western websites, as well as its Golden Shield, which carries out online surveillance, and its Great Cannon, which can be used to attack hostile websites. Microblogs and social networks such as Sina Weibo are policed aggressively, with prison sentences for those convicted of posting false or subversive information online. In September 2016, to give just one example of how the authorities operate, NetEase was forced by the government to close down all of its online forums, except for those on real estate and home.[63]

Yet censorship is not the key to the Chinese response to the networked age. The core of the strategy has been, by fair means and foul, to limit the access of the big American IT companies to the Chinese market and to encourage local entrepreneurs to build a Chinese answer to FANG. While Yahoo and Microsoft accepted government mandated "self-discipline," Google pulled out of China in 2010 after repeated wrangles with the Chinese authorities over censorship and attacks on human rights activists' Gmail accounts.[64] Ever since it registered the domain name www.facebook.cn in 2005, Facebook has tried to establish itself in China, but it was blocked in 2009, when Western social media companies were accused of fomenting unrest in mainly Muslim Xinjiang.[65] The result is that the internet in China today is dominated by BAT: Baidu (the search engine, founded by Robin Li in 2000), Alibaba (Jack Ma's answer to Amazon, founded in 1999), and Tencent (created the year before by Ma Huateng, best known for its WeChat messaging app). These conglomerates are much more than clones of their US counterparts. Each has shown itself to be innovative in its own right— and with a combined market value in excess of $473 billion and annual revenues of $20 billion, they are almost as large in scale as their US counterparts. WeChat is used by 86 percent of Chinese internet users and is fast replacing the once mandatory Asian business card with easy-to-snap QR codes. Alibaba's revenue in China exceeded Amazon's in the United States in 2015; its share of total retail revenue in China (over 6 percent) is twice that of Amazon's in the United States.[66]

Needless to say, Silicon Valley gnashes its fangs at being shut out of the vast Chinese market. Zuckerberg has not yet abandoned hope, giving interviews in fluent Mandarin and even jogging through the smog of Tiananmen Square, but the recent experience of Uber cannot encourage him. Last year, after incurring losses in excess of $1 billion a year, Uber ran up the white flag, accepting that it could not beat the homegrown ride-sharing business Didi Chuxing.[67] This outcome was a result partly of Didi's great agility and deeper pockets, but partly also of regulatory changes that seemed designed to put Uber at a disadvantage in the Chinese market.[68] The frustration of Silicon Valley with these setbacks is understandable. Yet it is hard not to be impressed by the way China took on Silicon Valley and won. It was not only smart economically; it was smart politically and strategically, too. In Beijing, Big Brother now has the big data he needs to keep very close tabs on Chinese netizens. Meanwhile, if the NSA wants to collect metadata from the Middle Kingdom, it has to get past the Great Firewall of China.

The conventional wisdom in the West remains that the networked age is as inimical to the rule of the Chinese Communist Party as it was to the Soviet Union. But there are those who beg to differ.[69] For one thing, the CCP itself is a sophisticated network in which nodes are interconnected by edges of patronage and peer or coworker association. On the basis of between-ness centrality, for example, Xi Jinping is as powerful as any leader since Jiang Zemin, and much more powerful than Deng Xiaoping, with whom he is sometimes wrongly compared by Western commentators.[70] Network analysis is allowing students of Chinese government to move away from simplistic theories about factions and to realize the subtlety of modern *guanxi*. Cheng Li has emphasized the importance of mentor-protégé ties in Xi's ascent to power—those relationships between senior party figures and their right-hand men (*mishu*). Those who distinguish between an elitist "Jiang-Xi camp" and the populist "Hu-Li camp" are exaggerating the rigidities of faction. Xi himself rose from being secretary to the minister of defense, Geng Biao, to later hold county-level and provincial positions in Hebei, Fujian, Zhejiang, and Shanghai, where he built up his own network of protégés, including figures as different as the "economic technocrat" Liu He and the "conservative military hawk" Liu Yuan.[71] As Franziska Keller argues, China is better understood in terms of such networks of mentorship than in terms of factions (see figure 6.3). Other important networks include the one formed by members of Xi's leading small groups and the one connecting corporations to banks via the bond market.[72]

Far from wanting to nail Jell-O to the wall, the Chinese approach to social media is increasingly to take advantage of what microblogs reveal about citizens' concerns. When researchers from Hong Kong, Sweden, and the United States mined a dataset of more than thirteen billion blog posts on Sina Weibo between 2009 and 2013, they were surprised to find that 382,000 posts alluded to social conflicts and as many as 2.5 million mentioned mass protests such as strikes. The hypothesis is that the authorities are now using social media to monitor dissent as well as to police corruption. Significantly, of 680 officials accused of corruption on Weibo, those eventually charged were mentioned nearly ten times more often than those not charged.[73] Another dataset—of 1,460 officials

FIGURE 6.3   The Chinese Communist Party Central Committee Members' Network

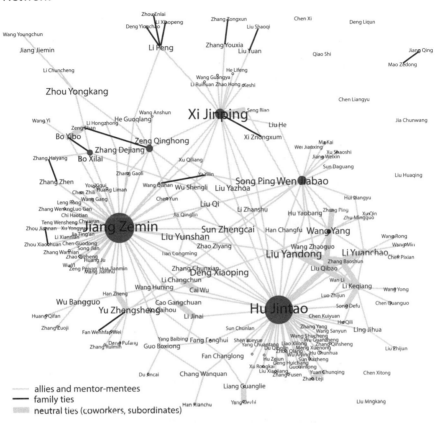

The size of the node is proportional to the number of connections (degree); the size of the name proportional to between-ness centrality. Note how ties between mentors and mentees matter much more than family ties.

Source: Franziska Barbara Keller, "Moving Beyond Factions: Using Social Network Analysis to Uncover Patronage Networks among Chinese Elites," *Journal of East Asian Studies* 16, no. 1 (May 2016).

investigated for corruption between 2010 and 2015—provides a further insight into the networks that run China, in this case the network of "tigers and flies" (i.e., big and small offenders) whose misconduct has become a key target of Xi Jinping's government.[74] The possibility exists that information and computer technology could enable Beijing to build a system of "social credit," analogous to financial credit in the West, that would (in official parlance) "allow the trustworthy to roam everywhere under heaven while making it hard for the discredited to take a single step."[75] China already has established systems of *hukou* (household registration) and *dang'an* (personal records), as well as schemes for rewarding outstanding workers and party cadres. Integrating these with the data that the authorities can easily glean from the BAT companies would provide a system of social control beyond the dreams of the mid-twentieth-century totalitarian states.

At the same time, China's leaders seem much more adept in webcraft than their American counterparts. While the Trans Pacific Partnership may be revived, but without the United States as a member, Chinese initiatives such as the Belt and Road scheme and the Asian Infrastructure Investment are steadily attracting new participants. A fascinating test of the Chinese approach will be how far they are able to leapfrog ahead of the United States in the rapidly growing sector of financial technology. Since ancient times, states have exploited their ability to monopolize the issuance of currency, whether coins stamped with the king's likeness, banknotes depicting past presidents, or electronic entries on a screen. However, blockchain-based digital currencies such as bitcoin and ethereum offer many advantages over a fiat currency like the US dollar or the Chinese yuan. As a means of payment—especially for online transactions—a digital currency has the potential to be faster and cheaper than a credit card or wire transfer. As a store of value, bitcoin has many of the key attributes of gold, notably finite supply. As a unit of account, to be sure, it is less than stable, but that is because it has become an attractive speculative object. Worse, bitcoin seems extraordinarily wasteful of computer resources because of the way that it is "mined" or "hashed" and authenticated.[76] On the other hand, bitcoin's distributed ledger technology appears to solve the problem of authentication and security so well that bitcoin can also function as a fraud-proof messaging technology, while ethereum can even automate the enforcement of contracts without the need for the bureaucratic monitoring that is an integral and expensive part of the existing system of national and international payments.[77] In short, "trust is distributed, personalized, socialized . . . without the need for a central institution for verification."[78] Of course, the Chinese authorities are no more ready to hand their payments system over to bitcoin than they were to hand their taxi system over to Uber. Indeed, alarmed that Chinese "miners" accounted for 40 percent of the global bitcoin network, with close to three-quarters of bitcoin trades occurring on the BTCC (Bitcoin China) exchange, financial regulators announced a regulatory tightening on the sector late in the summer of 2017. Within one month, the major privately operated Chinese exchanges "voluntarily" ceased domestic operations. However, Beijing clearly appreciates the potential of block-chain as a technology. That is why the People's Bank of China and a number of provincial governments are close to launching an "official crypto-currency"—"bityuan," perhaps—in one or two provinces in the near future.[79] If these experiments are successful, it would represent the beginning of a new epoch in monetary history and a serious challenge to the dollar's future as the principal international currency.

We live in a strategy-free world with respect to national security, and I think in many respects we live in a strategy-free world with geoeconomic issues, too.   —John B. Taylor

---

* * *

At times, it seems as if we are condemned to try to understand our own time with conceptual frameworks more than half a century old. Since the financial crisis, many economists have been reduced to recycling the ideas of John Maynard Keynes, who died in 1946. Confronted with populism, writers on American and European politics repeatedly con-fuse it with fascism, as if the era of the world wars is the only history they have ever studied. Analysts of international relations seem to be stuck with terminology that dates from roughly the same period: realism or idealism, containment or appeasement, deterrence or disarmament. George Kennan's "Long Telegram" was dispatched just two months before Keynes's death; Hugh Trevor Roper's *Last Days of Hitler* was published the following year. Philip Zelikow hankers after a new Marshall Plan.

Yet all this was seventy years ago. Our own era is profoundly different from the mid-twentieth century. The near-autarkic, commanding and controlling states that emerged from the Depression, World War II, and the early Cold War exist today, if at all, only as pale shadows of their former selves. The bureaucracies and party machines that ran them are defunct or in decay. The so-called administrative state is their final incarnation. Today, the combination of technological innovation and international economic integration has created entirely new forms of network—ranging from the criminal underworld to the rarefied "overworld" of the World Economic Forum—that were scarcely dreamed of by Keynes, Kennan, or Trevor Roper.

Winston Churchill famously observed, "The longer you can look back, the farther you can look forward." We, too, must look back longer and ask ourselves the question: Is our age likely to repeat the experience of the period after 1500, when the printing revolution unleashed wave after wave of revolution?[80] Will the new networks liberate us from the shackles of the administrative state as the revolutionary networks of the sixteenth, seventeenth, and eighteenth centuries freed our ancestors from the shackles of spiritual and temporal hierarchy? Or will the established hierarchies of our time succeed more quickly than their imperial predecessors in co-opting the networks and enlisting them in their ancient vice of waging war?

A libertarian utopia of free and equal netizens—all interconnected, sharing all available data with maximum transparency and minimal privacy settings—has a certain appeal, especially to the young. It is romantic to imagine these netizens spontaneously rising up against the world's corrupt elites, then unleashing the might of artificial intelligence to liberate themselves from the drudgery of work, too. Those who try to look forward without looking back very easily fall into the trap of wishful thinking. Since the mid-1990s, computer scientists and others have fantasized about the possibility of a "global brain"—a self-organizing "planetary superorganism."[81] In 1997 Michael Dertouzos looked forward to an era of "computer-aided peace."[82] "New information technologies open up new vistas of non-zero sumness," wrote one enthusiast in 2000. Governments that did not react swiftly by decentralizing would be "swiftly . . . punished."[83] N. Katherine Hayles was almost euphoric. "As inhabitants of globally interconnected networks," she wrote in 2006, "we are joined in a dynamic co-evolutionary spiral with intelligent machines as well as with the other biological species with whom we share the planet." This virtuous upward spiral would ultimately produce a new "cognisphere."[84] Three years later, Ian Tomlin envisioned "infinite forms of federations between people . . . that overlook . . . differences in religion and culture to deliver the global compassion and cooperation that is vital to the survival of the planet."[85] "The social instincts of humans to meet and share ideas," he declared, "might one day be the single thing that saves our race from its own self destruction."[86] "Informatization," wrote another author, would be the third wave of globalization.[87] "Web 3.0" would produce "a contemporary version of a 'Cambrian explosion'" and act as "the power-steering for our collective intelligence."[88]

Histories of futurology give us little reason to expect much, if any, of this to come true. Certainly, if Moore's Law continues to hold, computers should be able to simulate the human brain by around 2030. But why would we expect this to have the sort of utopian outcomes imagined in the preceding paragraph? Moore's Law has been in operation at the earliest since Charles Babbage's "Analytical Engine" was (partly) built before his death in 1871, and certainly since World War II. It cannot be said that there has been commensurate exponential improvement in our productivity, much less our moral conduct as a species. There is a powerful case to be made that the innovations of the earlier industrial revolutions were of more benefit to mankind than the most recent one.[89] And if the principal consequence of advanced robotics and artificial intelligence really is going to be large-scale unemployment, the chances are surely quite low that a majority of mankind will uncomplainingly devote themselves to harmless leisure pursuits in return for some modest basic income.[90] Only the sedative-based totalitarianism imagined by Aldous Huxley would make such a social arrangement viable.[91] A more likely outcome is a repeat of the violent upheavals that ultimately plunged the last great networked age into the chaos that was the French Revolution.[92]

Moreover, the suspicion cannot be dismissed that, despite all the hype of the networked age, less benign forces have already learned how to use and abuse the "cognisphere" to their advantage. In practice, the internet depends for its operation on submarine cables, fiber-optic wires, satellite links, and enormous warehouses full of servers. There is nothing utopian about the ownership of that infrastructure, nor the oligopolistic arrangements that make ownership of major web platforms so profitable. Vast new networks have been made possible but, like the networks of the past, they are hierarchical in structure, with small numbers of super-connected hubs towering over the mass of sparsely connected nodes. And it is no longer a mere possibility that this network can be instrumentalized by corrupt oligarchs or religious fanatics to wage a new and unpredictable kind of war in cyberspace. That war has commenced. Nor can it be ruled out that a "planetary superorganism" created by the Dr. Strangeloves of artificial intelligence may one day run amok, calculating—not incorrectly—that the human race is by far the biggest threat to the long-run survival of the planet itself and exterminating the lot of us.[93]

The lesson of history is that trusting in networks to run the world is a recipe for anarchy: at best, power ends up in the hands of the Illuminati, but more likely it ends up in the hands of the Jacobins. Some today are tempted to give at least "two cheers for anarchism."[94] Those who lived through the wars of the 1790s and 1800s learned an important lesson that we would do well to relearn: unless one wishes to reap one revolutionary whirlwind after another, it is better, for the sake of order, to impose some kind of hierarchical order on the world and to give it some legitimacy.

At the Congress of Vienna, the five great powers agreed to establish such an order, and the pentarchy they formed provided a remarkable stability for the better part of the century that followed. Just over two hundred years later, we confront the same choice they faced. Those who favor a world run by networks will end up not with the utopia of their dreams but with a world divided between FANG and BAT and prone to all the pathologies discussed above, in which malignant subnetworks exploit the opportunities of the World Wide Web to spread virus-like memes and mendacities.

The alternative is that another pentarchy of great powers recognizes their common interest in resisting the spread of jihadism, criminality, and cybervandalism, to say nothing of climate change and other shared threats. In the wake of the 2017 WannaCry episode, even the Russian government must understand that no state can hope to rule Cyberia for long: the malware was developed by the American NSA as a cyber-weapon called EternalBlue, but was stolen and leaked by a group calling itself the Shadow Brokers. It took a British researcher to find its "kill switch," but only after hundreds of thousands of computers had been infected, including American, British, Chinese, French, and Russian machines. What could better illustrate the common interest of the great powers in combating internet anarchy? They surely have as strong an incentive in this area as they do to combine against nuclear proliferation, the development and use of biological and chemical weapons, and the perils of pollution, not forgetting the spread of Islamic extremism.

Conveniently, the architects of the post-1945 order created the institutional basis for such a new pentarchy in the form of the permanent members of the UN Security Council, an institution that retains the all-important ingredient of legitimacy. Unlike the Marshall Plan—which the United States could implement with any willing partner on a bilateral basis—the UNSC was a failure in the sense that it did not remotely live up to the hopes of Franklin Roosevelt. The Soviet Union used its veto power the most often of the P5, casting eighty vetoes between 1946 and 1969. Between 1970 and 1991, however, the United States nearly equaled that total. In short, the idea of a hierarchy of powers, which had worked so well between 1815 and 1914, was unworkable in the polarized climate of the Cold War. Ideas that did work in the 1940s are probably of limited utility today; you need a current account surplus for a Marshall Plan. But ideas that did not work in the 1940s are another matter. Whether or not the five great powers can make common cause once again, as their predecessors did in the nineteenth century, is the great geopolitical question of our time.[95]

Reflecting on governance, and thinking back to my own experience in various jobs, I look over there at Sam Nunn and Bill Perry. Why did things seem to function relatively well in that era? I think one of the reasons is that we trusted each other. As we dealt with each other, if I said to Sam I would do something, I would go out of my way to be sure I did it, so he could be sure he could trust me. We made a deal.   —George P. Shultz

---

7

GOVERNANCE FROM AN INTERNATIONAL PERSPECTIVE

William Drozdiak

Disruptive technologies pose a pervasive challenge to democracies around the world. These engines of human progress bring vast benefits. But they also pose risks to fundamental values of democracy. In some cases, new technologies that offer a great leap forward for humanity—instant communications around the globe or the creation of driverless cars that curtail deadly accidents—can also be viewed as potential threats to political order, shared prosperity, and stable communities.

The internet was once regarded as a powerful weapon of democracy, an ingenious way to circumvent government censorship and empower the voices of common people. The Arab Spring rebellions that toppled dictators in Tunisia, Libya, Egypt, and Yemen heralded this new era. Yet now many autocracies have learned to employ digital tools to further suppress their people by blocking access to free information or by distorting the news that they receive. As they have in the past, autocracies claim such actions are justified to ensure the security of society over individual freedoms. They argue that in spite of rampant corruption and human rights abuses, the relative stability that prevails under autocrats is a better state of affairs than the violent anarchy that would follow their overthrow. Only now they can unleash new electronic tools of surveillance and repression of dissent. And as seen by Russia's meddling in Western elections, autocrats are also willing to distort information as a tool of hybrid warfare against democratic societies as a way to wage aggression without declaring war.

Criminal gangs have also learned how to exploit the internet. In February 2016, digital bank robbers infiltrated SWIFT, the global banking payments system, and stole $81 million from the central bank of Bangladesh, which was the largest hack ever of an international financial institution.[1] Later that year, Yahoo revealed a breach of more than one billion user accounts—about one-third of the global population of the internet at the time—in which passwords and encrypted security questions were stolen. And the Mirai botnet attack, which targeted household devices and electronic products, disabled the internet's domain name system and brought down sites including Twitter, Netflix, and CNN in an attack that was twice as powerful as any previous disruption on record.

The number of devices connected to the internet now exceeds our global human population of 7.5 billion. By 2020, according to the digital technology company Cisco, there will be more than fifty billion devices and sensors connected to an internet of things that will link smartphones, parking meters, thermostats, bank accounts, cardiac monitors, cars, supermarket shelves, and nearly every aspect of human life. The world has barely begun to grapple with the security and economic implications of building out this vast and vulnerable global network, involving billions of points of interconnection that could be disrupted by criminals, warlords, or mad dictators controlling a small army of hackers.

Our societies are becoming more diverse than ever, even as we wrestle with the forces of globalization. Successive waves of immigration have deepened our cultural, political, economic, and historical connections with the rest of the world. This growing diversity, coupled with the rise of digital social media, has created many fault lines and vulnerabilities within democracies that can be exploited by possible aggressors, whether they are based in Moscow, Beijing, Pyongyang, or Tehran, or have no fixed address, such as criminal syndicates.

Once-marginalized elements—from alt-right white supremacists in America to the Islamic State in the Middle East to hacker groups in Russia and Eastern Europe—can now rapidly cultivate a degree of social and political influence once deemed unthinkable. With about one-quarter of the world's population now on Facebook, the possibilities of extremist or hateful groups exploiting political grievances and cultural resentments in ways that cannot be controlled now threaten our democratic way of life.

Just as governments need to become more vigilant about unconventional forms of warfare, the public needs to be educated to become more discerning about the quality and credibility of information in the digital age. A promising development is the creation of popular news literacy courses for civil society that instruct young people how to navigate the maze of social media networks and to be discriminating in finding the truth by questioning the sources of what they read. Critical thinking skills have always been important, but they have become a vital necessity in the twenty-first century as people struggle to deal with information overload and to separate fact from fiction.

But fake news is only part of the story. The dangers posed by perverted or biased information are just a symptom of a larger truth now dawning on the world: with billions of people glued to Facebook, WhatsApp, WeChat, Instagram, Twitter, Weibo, and other popular services, social media is rapidly becoming perhaps the world's most dominant cultural and political force, to the point that its effects can alter the course of global events. Yet our democracies have not even come close to gaining a clear understanding of this explosive trend, let alone developing the means and methods to tame its power and ensure our values are protected.[2]

Political parties are being disrupted by radical reformers with no traditional power base who build their support online, as seen in the recent candidacies of Bernie Sanders in the United States and Emmanuel Macron in France. In the future, insurgent candidates will become more common as voters embrace politicians and policy prescriptions once viewed as outside the mainstream.

Besides the need to protect our political process and civic institutions from being disrupted by fake news and false information, the threat to our national security interests may be even more urgent. Following the failure to prevent Russian hackers from intervening in the election process, the United States needs to shore up its defenses against future intrusions. Rapid responses by Estonia, France, and Germany to thwart Russia's apparent efforts "to wage war without going to war" offer salutary lessons for the United States about the importance of stronger cyberdefenses in protecting vital infrastructure and raising vigilance levels during election campaigns. After neglecting the rising importance of cyberattacks in recent years, the NATO alliance now recognizes that "nonkinetic" weapons of hybrid warfare like hacking, subversion, espionage, and fake news pose serious threats to our security. Following Estonia's lead, Germany's new cybercommand was launched in April to help safeguard last September's election. Berlin's defense ministry has declared that cyberspace will now be considered the sixth branch of its armed forces.

The Western allies together need to develop a more sophisticated international strategy to block cyberattacks while still preserving democratic principles. NATO managed to win the Cold War with an effective deterrence strategy that contained Soviet aggression short of engaging in a nuclear conflict. As part of a new deterrent strategy to thwart cyberattacks, the Western alliance may need to consider whether it should deploy counterpropaganda programs against fake news and media manipulation by foreign countries in the way Voice of America and Radio Free Europe were used to counter *dezinformatsia* campaigns waged against the West by the Soviet Union.

During the Cold War, we had Radio Liberty and Radio Free Europe. All of the archival material came to our archive here at Hoover, so we had some people study it. And there were two questions. Did anything work? Number two, what did we learn? It was clear that everybody thought there really was a huge impact, and we learned a lot about how to do this effectively.—George P. Shultz

---

The Future of Employment and Social Order in Western Democracies

How Europe and the United States respond to the transformation of work and society by new technologies has already become an acute concern about how we can sustain our democracies in the twenty-first century. If, as recent research suggests, every industrial robot absorbs up to six jobs, that could mean up to six million jobs could be lost over the coming decade. This trend will also affect the problem of income inequality, since the poor and the young will suffer most. According to a White House economic report, there is more than an 80 percent chance that automation will take a job with an hourly wage below $20.

John Kasich, the Republican governor of Ohio, is one public official who worries about the impact of this rapid technological change on the health of our democracies if ever larger numbers of the poor, the young, and the less educated are thrown out of work. Speaking over dinner at the Munich Security Conference in early 2017, Kasich predicted that, by 2030, many truck drivers could find their jobs obsolete as driverless vehicles take over the roads.

"If truck drivers are one of the largest sources of jobs in my state, what do I or my successors need to do to prevent mass unemployment and all the social turmoil that goes with it?" Kasich asked. "If you think our society is polarized today, what will happen when you have an even greater disparity between rich and poor two decades from now? What happens when massive numbers of blue-collar workers have lost all hope for earning a decent living? It is a question that could make or break our democracy."[3]

Technological developments are forcing European politicians and business executives to reexamine their assumptions about labor markets, welfare benefits, and the now politically explosive subject of immigration. For example, Germany's interior ministry has for years warned in annual reports that because of its aging population and low birth rates the country should be prepared to accept as many as 400,000 migrants a year over the next quarter century to sustain the structure of its export-driven economy and generate enough income taxes to pay for a generous health and welfare system. But in February this year, the ministry scaled down that assessment to 300,000 immigrants per year over the next forty years. One factor was the rapid growth of automation and artificial intelligence and the reduction in the labor force they make possible.

But perhaps even more important was the rise of xenophobic nationalists who have gained popularity by arguing against an influx of Muslim or African immigrants. And foreign adversaries seem ready to seek to exploit these tensions through fake news and social media. During the exodus of Syrian war refugees into Europe, Russia helped disseminate rumors on right-wing websites about alleged rapes and other crimes by immigrants as a way of stoking internal political divisions in Germany and elsewhere.

Chancellor Angela Merkel cited Germany's shrinking population as a reason to open the country's borders to more than one million refugees fleeing Syria's civil war in 2015. The immigrants were initially welcomed into Germany as Merkel and many commentators hoped they could rejuvenate the country's population and fill many of the job vacancies created by Germany's booming economy. But efforts to integrate so many Syrian, Afghan, Albanian, and North African immigrants have proved more difficult than imagined. And the startling rise in support for the right-wing Alternative for Germany party, which rose as high as 15 percent in some polls after demanding that Germany shut its doors to immigrants, forced Merkel to backtrack and adopt a less tolerant approach. The chancellor's tougher stance reassured some of her more conservative supporters and helped deflate right-wing gains in the weeks before the September vote.

Indeed, some European governments are now placing a greater priority on ensuring that their existing populations are capable of adapting to new jobs in a changing global economy rather than recruiting workers from abroad. Denmark's center-right government, supported by the right-wing Danish People's Party, has strongly discouraged the influx of immigrants and refugees by threatening to confiscate jewelry and valuables worth more than $1,000 (ostensibly to pay for their food, lodging, and health care). At the same time, it has expanded efforts to help retrain and reeducate all Danish citizens who want to learn digital skills as a pathway to find better jobs and maintain unemployment levels at some of the lowest in Europe.

The Danish government seeks to ensure that as many people remain employed as possible in order to pay for a generous welfare state that offers free health service, up to five years of university education without cost, and subsidized care for children and the elderly. To persuade those without jobs and living on welfare stipends to return to the labor force as quickly as possible, the government makes extraordinary efforts to prepare its citizens to accept different kinds of employment and not focus on a lifelong career.

France's new president, Emmanuel Macron, has said his goal in seeking to cut his country's 10 percent unemployment rate (as high as 25 percent among young people) is to create a "Scandinavian-style economy" by reproducing many of the programs carried out in Denmark. And in the United States, former Democratic presidential candidate Bernie Sanders often declared during his campaign that his ideal vision for America would be to see his country become more like Denmark.

One out of four Danish workers changes jobs every year, more frequently than any labor force in the developed world. Employers are given great leeway in terms of hiring and firing workers to adapt to a highly competitive global economy—and worker training programs that keep adult citizens engaged in active work as often as possible are vital to sustain Denmark's enviable affluence. Denmark spends proportionately almost eighteen times as much as the United States on worker training, according to the Organisation for Economic Co-operation and Development.

Europe's worker training programs are considered much more advanced than in the United States, which may help the continent adapt better to the disruptions caused by the digital age. Germany's highly successful *Berufschule* program offers young people who do not go to university the chance to enroll in vocational training schools in which the weekly schedule involves three days of classroom work combined with two days gaining hands-on experience in a company. This enables Germany to adapt vocational training to the modern needs of industry so that students can go straight into jobs. Some community colleges in the United States have shown an interest in emulating the German model, but for cultural and other reasons the American programs have failed to make a big impact in helping young people who do not go to university to learn the advanced skills that will lead to well-paying, sustainable jobs.

Other countries in the European Union are also pursuing labor policies designed to boost local employment in ways that may reduce dependence on immigrants. In the case of Poland, the country's right-wing government has shut the door to virtually all Muslim immigrants but welcomed Ukrainians who share the same Catholic faith. Poland's paramount leader Jaroslaw Kaczynski has spread false reports in the media claiming immigrants from the Middle East and Africa would bring "parasites and strange diseases" to Poland and must be kept out. Meantime, Germany and Scandinavian countries have placed a new emphasis on language skills before immigrants can expect to acquire good jobs.

Some European politicians believe the era of high-tech transformation will require democratic societies to think beyond questions like whether to accept more immigrants or how to fund the welfare state. They say that encouraging the moral benefits of work is critical to sustaining a healthy society—something that will become a vital issue for democracies as technology eliminates more jobs.

Former French prime minister Michel Rocard, who spent his last years promoting an education revolution from his seat in the European Parliament, became convinced that the emergence of digital technologies should require universities to become lifelong bastions of learning and thinking for people of all ages, not just the young.

Rocard, who died in 2016, believed it should become customary in Western democracies for people to return to universities every two decades to reinvent themselves and create new careers that would last ten to twenty years. "Nobody should think of spending forty years at a single trade or profession and then go into retirement," he told me. "You should engage in phases of study and reflection in your twenties, in your forties, and again in your sixties, in order to prepare yourself for two, three, or four different careers."

The global economy will change so fast under the influence of new technologies, Rocard predicted, that people will need to go back to school regularly in order to learn new skills, or they will be left behind, leading to greater political and social alienation in Western society. On that score, his forecast is proving accurate.

There's a segregation going on; globalization is retreating. But now we're seeing there are important reasons why we need to have ways of getting together and talking on a global basis.—George P. Shultz

---

America and the World: The Clash over Data Privacy

The spectacular success of America's social media giants is often cited abroad as a principal reason why the United States remains the dominant force in the global economy. The powerful forces of innovation and entrepreneurial energy in Silicon Valley are both admired and feared in Europe, which is often accused of trying to obstruct the growth of American companies and protect the interests of the continent's homegrown enterprises.

The European Union has asked Ireland to seek $13 billion in back taxes from Apple, claiming it has unfairly exploited Ireland's low corporate tax rate while doing business in the twenty-eight-nation union. Apple has vowed to fight the EU demand. Airbnb is another company that has attracted the ire of European governments for escaping high taxes. Even though it has more than ten million users in France, Airbnb paid less than 100,000 euros in taxes last year in France.

Finance Minister Bruno Le Maire has vowed that the new French government under Macron will join Germany in pushing for the European Union to set a new global standard in imposing much higher tax assessments against the American digital giants. "All companies need to pay their fair share in tax in all the countries they operate in," Le Maire said. "That's not the case today. It's time to change gear. . . . We want the EU to take the lead in tackling this global issue."[4]

The American tech giants complain they are being unfairly targeted because of their overwhelming success. They say that punishing their performance will stifle innovation and ultimately hurt the quality of products offered to local users. Such actions, they claim, are designed to prop up weaker local rivals to the detriment of the consumer and amount to a new form of European protectionism.

But there are deeper forces at work in Europe besides competitive jealousy. Europe is genuinely worried that the untrammeled influence of American social media giants could shatter their cultural traditions of data privacy, encourage the spread of libelous rumors or "fake news," and generally erode civic democratic values. The horrible history of totalitarian abuses under Nazi, Fascist, and Communist regimes during the past century has made Europeans much more sensitive to the vast accumulation of personal data in the hands of the state or corporations.

Rather than seeing social media as a liberating force, Europeans tend to look toward the dark side and see a dystopian vision that could eventually dominate many aspects of human activity. "For us, big data equals big brother," a prominent Central European politician told me. "In the communist days, we used to worry about informers telling the government about the details of our private lives. Yet now, we inform on ourselves to the world at large through Facebook and Google."

Such distrust is spreading in Europe and making policy-makers uneasy about allowing social media to acquire enormous power and influence under the guise that free speech must remain a fundamental right in any democracy. Many Europeans feel that limits on hate speech, child pornography, and terrorist propaganda may be necessary in a modern democracy to find the right balance between freedom and security.

"Social media can challenge the basic principles of democratic life," said Margrethe Vestager, the EU's antitrust chief who has waged fierce battles against Facebook, Google, Apple, and other social media titans.[5] "If we are not careful, social media could let us down. Because despite all the connections it allows us to make, social media can also lock us up in our own worlds. . . . And we can't have an open debate from inside separate worlds."

"Lately, politicians have been learning a lesson that business has known for a long time," Vestager told a Brussels conference about democracy in the digital age. "The information that social media companies collect about their users can transform the way you advertise. It can help you put your message in front of exactly the people who are likely to buy it. But when you apply that to politics, it could undermine our democracy."

As one of Europe's most influential policy-makers, Vestager said she believes that the digital age presents serious challenges to basic rights such as personal privacy. She has been instrumental in pushing through new rules on data privacy across the European Union that will take effect in 2018 and are designed to protect the EU's five hundred million citizens from unwelcome intrusions that can be exploited by business.

She said, "More than four-fifths of Europeans feel that they don't have control over their personal information online. So we need rules to give them back that control."[6] She said an important part of new EU legislation is "data protection by design"—a principle that means "when you come up with a new digital service, you have to think from the start about how to protect people's privacy, so that treating people fairly isn't just an afterthought."

Many politicians across Europe share Vestager's view that digital technology companies must be held accountable for any unwanted exploitation of personal information. The United States has favored a softer approach, reflecting an aversion to putting too many limits on free enterprise and free speech. But in Europe, as in Asian democracies like South Korea and Japan, there is a greater willingness for the state to enforce tighter controls on abuses, such as hate speech and fake news, including the imposition of harsh penalties against companies that convey such information on internet platforms.

Vestager is reshaping antitrust law for the digital age. This year, she slapped a 2.42 billion euro penalty ($2.76 billion) on Google for abusing its dominant position in Europe as a search engine. The European Union also approved a pan-European digital privacy law that could have profound effects on the way social media companies use personal information. After four years of negotiations, the EU agreed to a common legal approach protecting digital privacy rights that will empower EU regulators to impose fines up to 4 percent of a company's worldwide revenue. This could lead to record penalties against American social media giants and other large corporations that have become data-dependent in the way they conduct business.

The purpose of the European law is to invest greater power in the hands of consumers and force large companies to respect their wishes when it comes to data-mining practices. The law will require these companies to seek additional consent every time they want to use such information. Online advertisers and data analytics firms say their business could be devastated and the EU approach will hurt innovation. They claim it shifts the burden of proof and could drag their companies into constant litigation to prove they are not at fault in any privacy violations. The law will also enshrine the controversial "right to be forgotten," which allows people to request deletion of personal data from online platforms like Facebook or Google.[7]

The EU claims it is upholding basic democratic values by protecting the rights of its citizens to control the most personal aspects of their lives. Privacy advocates believe the EU approach may soon become a model for the rest of the world. As Japan, South Korea, and other Asian nations consider adopting the tougher EU standards on data privacy, American companies will need to accept those rules or risk being excluded from lucrative markets. And the United States could find that it may have to adapt its own privacy laws in order to conform to the new global standard, at least among free-market democracies. US civil liberties advocates see digital privacy as a growing issue in the United States, and the European experience may be relevant here. Meantime, American companies will need to abide by Europe's judgments or find themselves hit with huge penalties that will hurt their reputation and ultimately erode their market presence there.

Some countries are taking even stricter measures to shield their citizens from growing abuses in the internet age. In Germany, where privacy fears are particularly acute because of repressive surveillance practices by the Nazi and Communist regimes, the federal parliament has passed legislation making social media companies responsible for eliminating objectionable content that is posted online. As of October 1, 2017, companies like Facebook and Google can be subject to penalties up to fifty million euros ($57 million) if they fail to delete within twenty-four hours any material construed to be hate speech, libel, terrorist propaganda, or other content deemed as "clearly illegal" by German authorities.

The new German law was approved months ahead of the September 24, 2017, federal elections, just as the anti-immigrant Alternative for Germany party had surged to more than 15 percent support—well above the threshold to qualify for seats in the federal parliament. A spate of xenophobic threats against immigrants carried on right-wing websites was believed to have encouraged hostile acts that included burning down housing for refugees. In addition, German authorities detected a sharp increase in online terrorist recruitment sites, which they feared could radicalize some of the more than one million refugees, many of them young Arab males from Syria and Iraq who have settled in Germany since 2015.

German police warned that an increasing number of false online reports were inciting hostile acts against immigrants. A Breitbart News report carried on the German internet claimed more than one thousand young immigrants had attacked police in Dortmund, waving Islamic State and al-Qaeda flags and setting fire to Germany's oldest church. In fact, a few men had simply set off firecrackers to celebrate New Year's Eve, there were no attacks on police, and a small fire on a piece of scaffolding had been quickly extinguished.[8]

Social media companies and civil rights groups claimed the new German law could suppress freedom of speech and be exploited by authoritarian regimes to justify their crackdowns on political opponents.

It could also stifle innovation to such an extent that many companies may question whether the risks of litigation or huge fines are worth the effort of doing business in Europe.

Anders Ansip, a former prime minister of Estonia who now serves as the EU's digital affairs chief, says the growing plague of fake news and its impact on elections in Western democracies drove the European Union to take action that would compel social media companies to assume greater responsibility in policing the internet. He denies that Europe's governments are engaging in de facto censorship of online content by threatening to inflict heavy penalties against companies that do not take active measures.

As somebody who grew up under communism, Ansip said that while some limits are necessary, he has no desire to see Orwellian media controls imposed on digital technologies introduced in Europe. "Fake news is bad but the ministry of truth is even worse," he said.[9]

A trendy new discussion is that this is the new Gilded Age and one has to "break up the Carnegie Steel Corporation." This is the wrong analogy. . . . Breaking up Google or Facebook would be a futile endeavor because as network economics predicts, these are pretty natural monopolies that have emerged. The real issue is that if social network platforms have become broadcast networks or publishers or giant media groups, they need to be regulated as such.   —Niall Ferguson

---

Disinformation, Cyberattacks, and Election Meddling

The revelations that Russia was involved in multiple efforts to influence the 2016 American presidential election in favor of Donald Trump came as no surprise to many Europeans. Indeed, despite the intense spotlight focused on Russia's interference in the 2016 US presidential election, Europe has become arguably the world's most active battle-ground in modern cyberwarfare. Since 2014, many European governments, including Germany, France, Latvia, Estonia, Sweden, and Montenegro, have been subjected to waves of mysterious cyberattacks and malicious falsehoods spread through the internet by what is presumed to be a Russia-directed onslaught.

Last January, a report released by US intelligence agencies confirmed what many people across the continent already assumed: Russia was actively seeking to influence elections across Europe, in what appears to be a much larger strategy of Russian covert actions designed to destabilize Western democracies. At the Munich Security Conference a month later, Russian foreign minister Sergey Lavrov, while denying any interference in US or European elections, spelled out Russia's ambitions to create conditions for what he described as "a post-Western era."

Russia's blueprint for a cyberwarfare strategy against the West was first outlined in a 2013 article in a professional military journal by General Valery Gerasimov, chief of staff for Russia's General Staff and a close adviser to President Vladimir Putin. Gerasimov claimed the huge military superiority of the United States could be effectively countered in cyberspace, which he said "opens wide asymmetrical possibilities for reducing the fighting potential of the enemy."[10]

He said Russia should learn lessons from the Arab Spring, when social media played a key role in mobilizing protests that brought down several entrenched dictatorships across North Africa and the Middle East. "We witnessed the use of technologies for influencing state structures and the population with the help of information networks," Gerasimov wrote. "It is necessary to perfect activities in the information space, including the defense of our own objects."

Gerasimov emphasized that Russia's military services needed to hone their hacking skills to serve as a surreptitious extension of conventional warfare and political conflict. He suggested that the use of disinformation, hacking, and deception through social media channels would be a way "to fight a war without fighting a war" against the West. Since the article appeared, Russia has greatly accelerated its political warfare campaign against Western democracies.

Last February, Russian defense minister Sergei Shoigu confirmed the existence of "information troops" that had long been denied by Moscow. "Propaganda must be smart, literate, and effective," Shoigu told Russia's lower house of parliament. According to the *Kommersant* business newspaper, Russia's military devotes about $300 million a year to a "cyber army" of about one thousand highly trained hackers. At their Warsaw summit in July 2016, NATO leaders adopted a Cyber Defense Pledge and vowed to make a top priority of protecting their digital networks and infrastructure. Since then, Britain has announced it would invest more than $2 billion in a national cybersecurity program and France has dedicated more than $1 billion to upgrading its cyberdefenses.

While Russia has engaged in spreading propaganda for decades, its intelligence services have greatly escalated their conflict with the West within the past three years by deploying the tools of digital technology in the ways that Gerasimov proposed. Kremlin specialists say Moscow's cyberoffensive against the West was likely triggered by the Maidan revolution in Ukraine, which Putin viewed as a Western-orchestrated effort that posed a direct threat to his own regime.

One of Russia's most important priorities has been Germany. Chancellor Merkel's staunch criticism of Russia's annexation of Crimea and its armed support for separatists in eastern Ukraine as well as her dominant leadership role in rallying support across Europe for sanctions against Moscow have made her government a primary target for cyber-destabilization efforts.

In December 2016, the German government informed the federal parliament, or Bundestag, that computer networks were being struck at least once a week by foreign intelligence services, mostly Russian. The Interior Ministry said in its annual report on security threats that "it is assumed that Russian state agencies are trying to influence parties, politicians and public opinion, with a particular eye to the [September 24] 2017 election."

Hans-Georg Maassen, the head of Germany's domestic intelligence service, said the intention might be "to damage trust in and the functioning of our democracy so our government should have domestic political difficulties and not be as free to act in its foreign policy as it is today."[11] He said it was initially assumed that Russia wanted to help Donald Trump in the US presidential election and that the cyberattacks were now seen as having "damaged American democracy."

Maassen said his agency was convinced that Russia would escalate its intrusions over the course of Germany's election campaign. He said a massive theft of electronic data from the Bundestag occurred during a 2015 cyberattack and that material might be released via WikiLeaks or other conduits before the September election. German officials believe the operation was carried out by the Russian hack group APT 28, also known as Pawn Storm, Fancy Bear, Sofacy, or Strontium. The hackers are believed to be controlled by Russia's military intelligence arm, or GRU.

The Russian government denies any connection to the hackers, but experts say there are hundreds of past incidents and suspicious connections that point to Russia. APT—for advanced persistent threat—was first identified by the global cybersecurity company FireEye. APT 28 has developed masterful "phishing" methods, using sophisticated fake emails with realistic but infected attachments that implant malware in foreign networks and can provide access to classified or sensitive materials.[12] These tools were used in the hacking of the emails of the Democratic National Committee ahead of the American presidential election in 2016. German officials claim that sensitive documents on US-German intelligence cooperation, presumably obtained through the Bundestag hack, were also published on WikiLeaks.

German security officials say the cyberattack on the Bundestag, which targeted the parliament's intelligence control committee, triggered the creation of a new cyberprotection department with more than ten thousand operatives to be run out of the defense ministry. Merkel also ordered a complete overhaul of the parliament's computer systems and deployed more sophisticated defenses to thwart any future cyberattacks aimed at sabotaging government institutions or key utilities such as power plants.[13]

Merkel's government was outraged by Russia's manipulation of the so-called Lisa case in 2016, when reports circulated about a thirteen-year-old Russian-German girl named Lisa who had been missing for two days and was allegedly raped by three refugees. The German police quickly learned the girl had not been raped and was merely staying with friends. But pro-Russian media sites kept insisting the girl had been raped by refugees, stirring up right-wing protests in Germany against the influx of Syrian refugees. Russia's foreign minister even accused Merkel of staging a cover-up of the truth.

The chancellor knows Russia will continue to deploy its arsenal of social media weapons to exploit the West's democratic traditions like freedom of expression, in contrast to the suppression of dissent in Russia and other autocracies. Again, this is nothing new. In the 1980s, Soviet disinformation efforts designed to split the Western alliance sought to persuade Europeans that the CIA was responsible for inventing the AIDS virus as part of an American biological weapons program.

Besides the loss of sensitive data and manipulation of fake news, the German government believes new dangers include the planting of delayed-action malware that could trigger "silent, ticking digital time bombs" in government computers and the nation's critical infrastructure. "This now belongs to normal daily life. . . . We must learn how to manage this," Merkel said in ordering the complete overhaul of government computers. She also insisted that industry help devise more effective ways to prevent sabotage of power plants, electrical grids, and other key parts of the national infrastructure.

Germany is not alone in Europe in finding its political process under attack through new digital technologies. Social media has helped make extremist candidates and causes seem more plausible, and not just in the United States, where Trump's victory took pundits by surprise. In the Philippines, Rodrigo Duterte cultivated a vast army of online supporters to help him win the presidency even though the crude, tough-talking mayor was heavily outspent by his mainstream opponents. In Britain, the once unlikely cause of leaving the European Union won an outright majority of votes in the June 2016 referendum thanks to an effective mobilization of supporters on Facebook and other forms of social media.

Russia has consistently sought to exploit extremist groups on the right and left to destabilize Western democracies. Moscow is finding stronger resonance for its support of extremist messages in European politics, in part because of the expanding impact of social media. In Hungary, the once-marginal far-right extremist group Jobbik has emerged as the leading opposition party, forcing conservative Prime Minister Viktor Orban to move sharply to the right. Both Jobbik and Orban's ruling Fidesz party have cultivated close ties with Putin's Russia, causing consternation among NATO allies.

This year, Russia acquired control of a far-right website in Hungary called Hidfo, or The Bridgehead, that now operates from a server in Russia and provides a platform for Russian disinformation. Orban, who has befriended Putin and shares his scorn for liberal democracy, also accepted Moscow's offer of a $10 billion loan for Hungary to pay for construction by Russia of a nuclear power plant. Elsewhere, Russia has tried to disrupt the normal functions of democracy in Scandinavia by nurturing far-right groups such as Nordic Resistance, which has formed an alliance with the Russian Imperial Movement.[14]

In Central and Eastern Europe, Russia's ambition to secure political control of its periphery by exercising greater influence through the acquisition of local radio and television stations, newspapers, and social media sites recently helped elect pro-Moscow candidates in presidential elections held in Bulgaria and Moldova. Just months after joining NATO, Montenegro's government was besieged by a wave of cyberattacks, presumably as a consequence of becoming a member of the Western military alliance against Moscow's wishes. The attacks came after twenty people, including two Russians, were arrested and charged with planning a coup.

But Russia's most important priorities appear to be centered on the larger countries in Europe. Besides Germany, Moscow has been particularly active in Italy and France. In Italy, where Putin once enjoyed a close friendship with billionaire former prime minister Silvio Berlusconi, Russia has sought to build ties with the anti-immigrant Northern League, which strongly opposes sanctions against Russia and whose leader, Matteo Salvini, has paid numerous visits to Moscow.

Russia is also courting Italy's populist Five Star Movement, which was launched less than a decade ago as an online movement to promote transparency in government. It is now leading in polls as the country's most popular party with about 30 percent support and aspires to head the next government after national elections are held, probably in early 2018.

The Five Star Movement was cofounded in 2009 by the comedian Beppe Grillo and the internet publishing entrepreneur Gianroberto Casaleggio, who died of a brain tumor in 2016. They have attracted a lot of support from Italian voters disillusioned with government corruption by promising that major policies and programs will be subject to public approval through online referendums. Five Star describes frequent online votes as its preferred format for ensuring transparent democracy and direct citizen participation in government. Casaleggio's son, Davide, who has inherited his father's internet business, has further honed the movement's use of online tools to raise funds and recruit candidates for public office.

Part of Five Star's appeal is to engage people directly through its online tools in proposing and drafting legislation. Despite the movement's naïve populism, Five Star has capitalized on a new desire in Italy for greater citizen empowerment as a way to help bridge the gap of distrust that evolved between political parties and the citizens they aspire to help. In that sense, they have become the country's leading political party because they are perceived as a fresh hope of restoring democracy.

Casaleggio's company also controls several popular websites that often publish sensational and distorted reports found on Sputnik Italia, an Italian version of the Kremlin-backed website that espouses Putin's views on the world. Luigi di Maio, a likely prime minister if Five Star wins the next elections, rejects sympathy with Moscow and says the movement is aligned with neither Russia nor the United States. But some Five Star members of parliament back Italy's departure from NATO and endorse Moscow's views on issues like the Syrian civil war.[15]

Russia was also deeply involved in meddling with France's 2017 presidential election campaign, seeking to bolster support in favor of the right-wing National Front candidate, Marine Le Pen. The First Czech Russian Bank, now in bankruptcy, loaned about $10 million to Le Pen's party, and Putin met with her in Moscow during the campaign to burnish her foreign policy credentials. She, in turn, has advocated the lifting of all sanctions against Russia and promised to pursue a new policy of greater Western cooperation with the Kremlin.

Moscow tried repeatedly to sabotage the candidacy of Emmanuel Macron and his *En Marche!* political movement by spreading rumors that Macron was a closet homosexual and "an agent of the big American banking system" because of his past employment as an investment banker with Rothschild. Macron's campaign websites and computer networks were targeted frequently in the months ahead of the presidential election, with hundreds if not thousands of attacks coming from hackers believed to be located in Russia, according to Richard Ferrand, the national secretary of Macron's party.

On the eve of the French presidential election, Macron's campaign staff members discovered they had been struck again, this time by a massive and coordinated operation unlike any of the others. The digital attack involved a dump of campaign documents including emails and accounting records in the hours just before a legal prohibition on campaign communications went into effect.[16] Macron's digital campaign director, Mounir Mahjoubi, told journalists the hackers had mixed fake documents with authentic ones "to sow doubt and misinformation." He said the operation, coming just before France's most consequential election campaign in decades, was "clearly a matter of democratic destabilization, as was seen in the United States during the last presidential campaign."

> *How can government structures and decision-makers take advantage in a big data era of decision-support mechanisms that are provided by technology such as AI? How do we bring into the governance process the exploitation of these new capabilities? That's something that I think governments don't do very well today.*   —Christopher Stubbs

What Can the West Do to Thwart Cyberattacks?

Shortly after his election, President Macron welcomed Putin to a huge cultural exhibition held in Versailles celebrating French-Russian relations. During a joint press conference, with an uncomfortable Putin at his side, Macron described RT and Sputnik as "agents of influence which on several occasions spread fake news about me personally and my campaign through lies and propaganda." He noted with evident satisfaction that Russia's disinformation efforts had proved in vain, as he won a convincing 67 percent of the vote in the final round against Le Pen.

RT and Sputnik had repeatedly released false voter surveys during the campaign, claiming Macron was running well behind Le Pen and Francois Fillon, the former conservative prime minister who had also cultivated friendly relations with Putin. Mahjoubi told journalists the main objectives of Russia's state-funded media outlets were to foment uncertainty about the election and spread chaos as a way of diverting attention from Fillon's legal problems. Fillon was the one-time favorite in the presidential race, but his standing plummeted when he became embroiled in a corruption investigation after putting his wife and children on the parliamentary payroll for doing little or no work.

Mahjoubi believes the principal reason that Russian attempts failed to disrupt the French election was because the campaign had made meticulous preparations to defend against potential cyberattacks. Macron's campaign team and French government authorities had anticipated a

Russian onslaught, not just because of what they saw occur during the US presidential campaign. In 2015, a massive cyberattack nearly shut down the world's largest francophone broadcaster, TV5Monde. Only a last-minute intervention by a TV technician, who ripped wires from a targeted server, prevented the collapse of the network. The hackers sent false messages to make it appear that the Islamic State was behind the attack, but French technicians later traced its origins to APT 28, or Fancy Bear, which turned out to be the same address that carried out attacks against the Democratic National Committee in 2016.

The fact that Macron's team and French authorities had drawn lessons from previous attacks and were bracing for a fresh wave of cyberassaults clearly helped mitigate the damage. As early as October 2016, the French national cybersecurity agency summoned all political parties involved in the campaign to raise their awareness about the risks of manipulation and outside interference that could impinge on national sovereignty in such a high-stakes election. In the months before the May 2017 election, France's defense ministry created a cyber command center composed of 2,600 experts knowledgeable in the ways of repulsing hack attacks. France's painstaking defenses clearly paid off.

France's success in blocking Russia's hacking operations—aided by French laws that prohibited the media from disseminating information in the final forty-eight hours before the election—showed that while open, democratic societies may be vulnerable, they are not helpless in thwarting such attacks. An open society can encourage its public to be alert and informed about the nature and intentions of forces seeking to disrupt the proper functioning of democratic processes, like free elections. While the murky nature of cyberoperations makes it difficult to trace the identity of actual culprits, there are hopeful signs that Western countries may be able to develop a more effective shield against channels of information warfare.

Over the past decade, the Estonian government has developed some of the world's most advanced and sophisticated defenses in countering cyberattacks. In 2007, the small Baltic nation of 1.3 million people was overwhelmed by a spate of cyberattacks, believed to originate in Russia, that nearly crippled its banking system and digital infrastructure. The websites of leading newspapers, political parties, and government ministries were also disabled. The attacks came shortly after Estonia removed a Soviet-era memorial to World War II from the center of its capital.

In early 2017, NATO Secretary General Jens Stoltenberg declared that Western alliance networks were being subjected to an average of five hundred attacks a month, up 60 percent from the previous year. While the Russian government consistently denies meddling in foreign elections and waging cyberattacks against the West, many of the intrusions are believed by NATO intelligence experts to have been conducted by Russian hackers operating with tacit, if not active, support from Moscow.

Since the 2007 attacks, Estonia has transformed itself into a highly valued Western ally that has erected NATO's most advanced cybersecurity defenses.[17] A recent alliance joint exercise called Locked Shields 2017 took place in the Estonian capital of Tallinn with nine hundred security experts from across Europe and the United States. They were challenged to defend against simulated attacks such as hackers breaking into an air base's fueling system and fake news reports accusing NATO of developing drones with chemical weapons.

Estonia has encouraged direct involvement by the private sector in helping fortify the nation's cybersecurity defenses. Under former president Toomas Hendrik Ilves, Estonia set up a program that enlisted volunteers who donate their free time, much like a national guard, to learning how to protect the nation's digital infrastructure, including everything from online banking to electronic voting systems. The program has already been emulated in neighboring Latvia. In the United States, the state of Maryland is consulting with Estonia about setting up a cyberdefense unit within its own national guard.

Despite such efforts, Russia-directed hacking threats continue to escalate in both Estonia and Latvia, the two Baltic nations that feel most highly exposed to such intrusions, in part because they have large ethnic Russian populations. Early this year, Marko Mihkelson, chairman of the Estonian parliamentary foreign affairs committee, received a suspicious email, allegedly from NATO, offering a link to an analysis of a North Korean missile launch. He did not click on the link but alerted some of Estonia's top cyberexperts. They found, yet again, that the message contained the same malware as that used against the Democratic National Committee during last year's presidential campaign by APT 28, alias Fancy Bear.

Developing better defenses and making Russia aware that further attacks traced back to Moscow will result in serious damage to relations with the West may be the best formula for an immediate, effective Western deterrent. When Merkel met with Putin and demanded that he personally halt Moscow's campaign of falsehoods in the Lisa case or face further escalation of economic sanctions and permanent damage to Russia's relations with Germany, Putin finally backed off. Since Merkel's direct warning to Putin, Russia has ceased making inflammatory comments about the Lisa case.

And when the United States challenged China about its cyberespionage efforts in 2008, China deflected blame and insisted it had done nothing wrong. Later, as Western defenses became stronger, China realized an escalating crisis with the West and the likelihood of economic sanctions were not worth the risk of conducting further cyberattacks for military or commercial purposes.[18] The prospect of diminishing returns and growing costs persuaded Chinese president Xi Jinping to agree to halt Chinese commercial cyberespionage, first against the United States and then against the United Kingdom and all G20 nations.

These examples suggest that the shield of stronger defenses and the stick of harsh sanctions can help contain the danger of cyberattacks spinning out of control. Even if Russia persists in its belligerent attitude toward the West and its aggressive efforts to destabilize its neighbors, a combination of smarter and stronger Western cyberdefenses along with punitive retaliatory measures may form the most effective basis for deterrence in a new age of hybrid warfare.

The Chinese story on cyber industrial espionage is the one that gives me hope too because they really did go from being the poachers to being gamekeepers. They started to understand, "Hey, we have IP. We're starting to innovate. This stuff is going to start hurting us."   —Niall Ferguson

---

REFLECTIONS ON DISRUPTION: CYBERWEAPONS

Nicole Perlroth

Over the last few years, my research has focused on the black market for something called a zero day vulnerability, which a lot of people were not familiar with—or at least they weren't four years ago. These days people are more familiar with this market thanks to an Apple–FBI–Department of Justice case a little over a year ago, when the DOJ sought in court to force Apple to create a backdoor into an iPhone used by one of the gunmen in the San Bernardino, California, shootings. If you remember, the DOJ officials ended up actually dropping the case. They said, "No, thanks. We don't need your help anymore. We actually were able to find a way into the phone through an outside party."

That outside party had found what's called a "zero day," which is just an unknown vulnerability in hardware or software that has not been patched and functions as a kind of invisible backdoor that allowed the government to get into this iPhone, and potentially many, many others.

Those weapons—zero days—can be used as espionage tools or as access points to drop more malicious, destructive cyberweapons, and they exist in much of the widely used software we rely on today. Some of the most coveted, for example, are in Apple and Microsoft software. One zero day hole in that software can fetch a million dollars in some cases on the black market—where governments, including our own, are actively paying hackers to turn over holes in this software and agree never to disclose them to the software maker for patching (which would render a zero day useless).

For the past few years, I've taken a deep dive into this market, its origins, and how it was catalyzed by the US government. One of the things that has been interesting is that Stuxnet, for all its publicity, did not just hit the computers that control Iranian centrifuges. It hit computers in thirty-one countries—and opened other countries' eyes to what these cyberweapons can do, the possibilities of offensive capabilities in cyberspace. It also jump-started the appetite for other countries to acquire their own stockpiles of cyberweapons and the means to deploy them.

In cyber, the countries that have the advanced capabilities to do harm are generally thought to be "the original five"—the United States, Russia, Israel, China, and the United Kingdom. Then we also have "the other five" players that have high intent to do harm, but capabilities that range from rudimentary, junior high–level skills to semi-advanced. This group generally includes North Korea and Iran, the Islamic State, al-Qaeda, and perhaps the Taliban.

But the problem is that the always-tenuous gap is closing between countries that have low intent and high capability, and the second group of actors with high intent and low capability. The reason that gap is closing is in part because of the market for zero days and other tools that can be easily acquired on the growing black market. There are now companies out there that sell "press and play" capabilities, particularly for cybersurveillance, maybe not yet for cyberweapons, who are actively marketing their services to countries that have the willpower, but not the talent or skills, to do actual harm.

As an interesting anecdote, I went down to Argentina to meet with the hacker community, in large part because the Argentine population is highly educated but their access to technology is fairly limited because of trade restrictions. To get access to certain apps and games, Argentine youth must learn how to hack. So the country has a very interesting culture of hacking. Many of them are naturally quite good at finding these vulnerabilities and zero days. And because of the currency exchange rates, if they sell one zero day to a nation-state broker for five or six figures, they can live pretty large in the trendy Palermo neighborhood of Buenos Aires.

I asked one of these hackers, "Who won't you sell to? Will you only sell to 'good' Western governments?" They said, "Nicole, you cute thing, you have to realize that we're in Argentina. We're not necessarily your ally, and the last time we checked, the country that bombed another country into oblivion wasn't Iran or North Korea. So we'll just sell to whoever has the biggest bag of cash."

When we talk about a "strategy-free zone" and national security issues, one of the things that has happened just over the last year, which has been catastrophic, was the Shadow Brokers leaks. Some of the NSA's own zero day and cyber arsenal was stolen and dumped online by a group calling themselves the Shadow Brokers. By most accounts, the NSA still does not know who exactly the Shadow Brokers are—it believes they are a combination of an insider and a nation-state, almost certainly Russia—but once the NSA's tools were leaked online, they could be used by anyone.

In one case, they were picked up in a widespread attack by North Korean attackers, called "WannaCry." You may recall a few months ago there was an attack that suddenly froze computers with ransomware, and attackers demanded a ransom in bitcoin to unlock a compromised user's data. The attack hit major companies in Russia, the United States, and Europe, and even shut down hospitals in Britain.

A week or so later, we saw a similar attack, using the same NSA tools, on a Ukrainian payroll system that appeared to be a targeted attack by Russian hackers on Ukraine. But what the attack demonstrated was that we're now in a globalized system, where lots of international companies actually pay contractors in Ukraine as part of their day-to-day business. Suddenly you saw Merck, Maersk, even companies in Tasmania paralyzed—having been taken down by a Russian attack using leaked NSA tools.

* * *

What do we do about this?

There is not much good news to report in this arena, though what I call "the resistance movement" has started taking shape in Silicon Valley. Recently I spent a lot of time in the bowels of Google, which has started a program to pay hackers bounties for turning over vulnerabilities in Google code. Google is paying them sums of money that are not as high as what nation-states will pay hackers, but it's paying more on the front end to keep hackers from weaponizing their code. For example, typically a nation-state or nation-state broker will require hackers to demonstrate how their zero day can be "weaponized" into an espionage tool or cyberweapon. But that takes time for testing and development, often months. By paying hackers just for the vulnerability alone, Google makes sure the vulnerability gets fixed and patched before it ever gets weaponized, and saves hackers some time.

Google has also started a project called Project Zero. It culled some of the best employees on its security team and hired hackers who have turned over bugs to its own bug bounty program, and essentially charged them with a mission to go around the internet, finding vulnerabilities in widely used products and code, in an effort to get them patched. The program is a bright light on the defense side of things.

But the United States is woefully behind in defense. For every one person working on information assurance and resilience at the NSA, there are eighteen people working on how to exploit code for information collection. That is the gap we are dealing with.

As far as international cooperation, the problem is that the United States does not have much ground to stand on. The US government has been actively exploiting software and paying hackers to turn over gaping security holes in their products for two decades now. It is not going to stop looking for zero days or exploiting them anytime soon. And, God forbid, if a terrorist attack were to happen tomorrow, and it turned out we could have somehow stopped it by getting into someone's encrypted iPhone to read the planning messages ahead of time— well, critics would call the NSA or maybe even the US government negligent for not doing more to get into that phone or computer operating system.

Importantly, as likely users of these weapons ourselves, we Americans have early on lost our halo on this topic. This market has spread far beyond US borders, where people are willing to sell these vulnerabilities to nation-states like Russia, Iran, or North Korea that have a very different moral compass and strategic calculation for how they will be used. This is a global problem, and solutions are hard to come by. It is not clear whether any kind of international code of conduct would work, but my point in doing this research is to compel governments to at least stop pretending these programs do not exist and start having the necessary conversations.

What I found in the zero day market realm is that people are not having these discussions, in part because data are logically hard to come by. The second you talk about one of these vulnerabilities, it gets patched, and suddenly a good that was worth a million dollars is now worth nothing. The United States and other countries are pouring millions into these programs, and the last thing they want is to see their espionage diamonds turned to mud.

But I think that ultimately the best way forward is to get this out in the open, admit that this is something we're doing, admit that it has crawled far beyond our borders, and try to organize settings—like this one—to talk openly about what is an acceptable use of these vulnerabilities and what is not an acceptable use. Or at least, inside the United States, we need to talk about the fact that we are clearly not protecting these methods enough from groups, like the Shadow Brokers, who are dumping them online and allowing our adversaries to use them back against us.

* * *

Unfortunately, I think that extrapolating nuclear weapons deterrent strategies to the cyberdomain won't work. One problem is determining attribution for any attack, which is a technically hard problem to either do today or create network architectures over time that could help. But deterrence is deficient also in large part because we are operating on a very asymmetrical battlefield. Yes, the United States is still the most advanced when it comes to offensive capabilities. But we are also the most vulnerable, because we are the most hyperconnected. Meanwhile, an adversary like North Korea has a very weak connection to the internet and is nowhere near as vulnerable as we are to cyberattack. So it's a very asymmetrical situation we find ourselves in.

One of the other things I would point to is that North Korea and Iran are heavily investing in their own offensive capabilities because they know they'll never be able to match us with kinetic warfare. We are all focused on mushroom clouds at the moment, particularly when we talk about North Korea and Iran, but no one's really focused on the fact that North Korea spent the better part of the last five years planting software "implants" in South Korea's critical infrastructure in the event of a rainy day, or kinetic escalation. Researchers have found evidence that Pyongyang's hackers have been exploiting vulnerabilities in South Korean systems and implanting malicious code at South Korean banks, utilities, and major companies that are the equivalent of "logic bombs" that can be launched to shut down South Korean systems, wipe data, paralyze South Korea's economy, or, in the worst case scenario, turn its power off.

We don't know yet how good they are at some of those offensive capabilities, but we know they now have NSA capabilities in their arsenal, and we know they've been actively infiltrating these systems for the past five years, so that's the situation we now find ourselves in.

Of course it wouldn't be novel to use a cyberweapon to target physical infrastructure. We've all been doing it to each other for the last decade. In fact, Russia has been actively targeting US energy networks and US energy companies and, most recently, some nuclear plants—not at the production level yet, but at the employee level—with increasing frequency. Their ultimate goals are not known, but it doesn't look like they are out to steal trade secrets. It looks more like the type of attack where they're laying the eventual groundwork for a future attack. China has been caught breaking into the computerized systems we use to control our industrial control systems as well.

I have zero doubts that the United States is doing the same. So perhaps we have reached a détente. We're all so implanted in each other's systems that we know the minute we launch something, they'll launch it back, or vice versa, which brings us full circle to mutually assured destruction.

8

GOVERNANCE AND THE AMERICAN PRESIDENCY

David M. Kennedy

My subject is the American presidency. I will mention some specific presidents, but my main concern is not with individual personalities. It is rather with the nature and attributes of the *office* of the president: its place in American constitutional architecture, its operational functions in the machinery of American governance, and its relation to American political culture—in particular, the standing of the office in the public's eye and its relation to the organs of the media as they have evolved over the last two and one half centuries.

A near perfect storm of converging forces, some with deep roots in past American experience, some born of more recent history, has converged to wallop the American political system with cyclonic energy in the first years of the present century. They include the changing methods of recruiting and electing candidates for office, especially the presidency; the increasing scale, complexity, impersonality, and volatility of a postindustrial economy and the chronically unmet need for instruments of governance with authority and agility commensurate

Author's note: This chapter builds on remarks delivered at an October 6, 2016, lecture to the Stanford University Political Science Department's "Election 2016" course.

with those attributes; the speed and phenomenal fragmentation of modern communication technologies; the decay of traditional institutions, public as well as private, conspicuously including political parties; and the intersection of the otherwise benign American values of freedom of speech, freedom of choice, physical and social mobility, and homophily that has spawned a culture of distrust that pervades virtually every aspect of modern American life.

Taken together, those developments have enormously stressed the systems and practices of governance that served the public interest reasonably well in the republic's first two centuries. Whether those legacy institutions and behaviors will prove sufficiently resilient to survive the new century's upheavals is a distressingly open question. So too is the fate of the values—another name for shared premises and even shared mythologies—that have sustained the American experiment in democracy since the nation's birth.

Five days a week, I think that everything that has happened in the last year or so is within the normal range of American politics. And two days a week I think, "This must have been how it felt in the final years of the Roman Republic." —Niall Ferguson

---

* * *

The nation may have been conceived in 1776, but it was truly born only with the adoption of the Constitution in 1788. In particular, the constitutional character of the presidential office—that is, the method of electing the president, and the powers both granted and denied to the executive branch—was defined by the men who drafted the Constitution in Philadelphia in the summer of 1787. Its fundamental attributes remain to this day deeply indebted to their deliberations, for better or worse.

The presidency's place in the actual operations of the American system of governance, however, is another matter, because in the 230 years since the Constitution was framed, that system has evolved—in many ways dramatically—along with the society and the economy in which it is embedded. So too have the citizenry's conceptions of the president's role and their expectations of the scope and scale of the president's responsibilities.

The tension between the essentially static constitutional character of the presidency and the manifestly dynamic society, economy, and culture in which it is embedded—and especially the technologies, even more especially the communication technologies, that have emerged over the last century—will be the main focus of this chapter.

* * *

Let's begin with some numbers.

There have been forty-five presidencies, but only forty-four presidents, thanks to the peculiar way that Grover Cleveland's two nonconsecutive terms (1885–1889 and 1893–1897) are counted. All forty-five have been men. All but two have been white Protestants. (The exceptions are the Roman Catholic John F. Kennedy and the African American Barack Obama.) Only seventeen, barely a third of all presidents, have been elected to second terms, which might serve to remind that even the pomp of power awaits the inevitable hour, and perhaps to suggest something distinctive about stability and consistency in American governance.

Among US presidents, twenty-six were trained as lawyers, eighteen previously served in the House of Representatives, seventeen previously served as governors, sixteen previously served as senators, fourteen previously served as vice president, and nine had been generals. Just three—and not a particularly happy three, I'm afraid—could be described as having had careers as "businessmen": Herbert Hoover, Jimmy Carter, and Donald J. Trump.

Two states are tied for having produced the most presidents, at seven each: New York (from Martin Van Buren to Donald Trump) and Ohio (the first was a Virginia transplant, William Henry Harrison, and the last was Warren G. Harding). That Ohio, a midsize state, has had such a prominent place in president-providing is suggestive of the quirks and caprices that have long characterized political career trajectories in a society that has no deeply entrenched governing class. At one time, the Ohio presidential phenomenon inspired a revision of Shakespeare's dictum that "Some are born great, some achieve greatness, and some have greatness thrust upon them" to read that "Some are born great, some achieve greatness, and some come from Ohio."

Virginia is the next most presidentially prolific state with five (none since John Tyler), and then Massachusetts with four (John F. Kennedy was the last). Six have come from the trans-Mississippi West, starting with California's Herbert Hoover, and five more since WWII, reflecting the phenomenal energy and rising demographic, economic, and political prominence of the Western region since the mid-twentieth century—two Bushes and Lyndon Johnson from Texas, and Richard Nixon and Ronald Reagan from California.

Thanks to the peculiarities of the Electoral College (about which more below), five presidents have been elected without popular majorities— John Quincy Adams in 1824, Rutherford B. Hayes in 1876, Benjamin Harrison in 1888, George W. Bush in 2000, and Donald Trump in 2016.

Eight have died in office, four of them assassinated (Lincoln, Garfield, McKinley, and Kennedy). Two have been impeached (Andrew Johnson and Bill Clinton), and one (Richard Nixon) has resigned.

There have been two father-son dynasties (John and John Quincy Adams and Bushes 41 and 43) and one grandfather-grandson combination (William Henry and Benjamin Harrison).

* * *

But of all those numbers, here's the single numeric datum that is most significant: the president is just one of the 536 federally elected officials in Washington, DC. (For these purposes I am treating the president and vice president as a single, unified political unit.) It's worth repeating that the president is but one of 536 elected officials in Washington, DC. The others, of course, are the 100 members of the Senate and the 435 members of the House of Representatives.

That legendary chronicler of several twentieth-century presidential elections, Theodore White, once captured the essence of that 1-to-535 ratio when he wrote that "the supreme duty of the President is to protect us from each other's Congressmen."[1] White's characteristically facile quip in fact points to some profound and persistently problematic attributes of the presidency—indeed, problems with the entire American political structure and system.

Those framers in Philadelphia more than two centuries ago were not only revolutionaries who had recently fought and won the War of Independence. They were also intellectual revolutionaries and serious political innovators. When they were drafting the Constitution, among other accomplishments, they essentially *invented* the presidency. No real precedent for the office had existed in the British colonies. Colonial executive power did not lie in the hands of an elected official, but in those of a royally appointed (and usually royally resented) governor. Nor was there any precedent in the Articles of Confederation that formed the original governing charter for the new nation. The Articles, in fact, made no provision for an executive office of any kind.

So those Founding Fathers got quite creative at Independence Hall in that Philadelphia summer. Acknowledging the weaknesses of the Articles of Confederation, they wanted an effective executive. But, remembering the abuses of those much-disliked royal governors, and ferociously opposed to anything that even faintly resembled monarchical power, they also feared the concentration of executive authority. How to strike the balance?

The result was the famous (or infamous) system of "checks and balances" that American students learn about (or once upon a time learned about) in high school civics classes. Power was deliberately dispersed and divided; lines of authority were purposely plotted to intersect at multiple points. The Framers conferred on the president the power to make treaties and to staff and manage the offices of the executive branch—and simultaneously hedged that power by requiring the advice and consent of the Senate on treaties as well as on high-level executive appointments. (Executive appointments often require congressional confirmation down to the fifth level of authority in many cabinet departments. The United States consequently has a far larger class of political appointees and a far smaller class of professional civil servants than most other advanced democracies, and hence—compared, for example, with France—less technical expertise in government and more politically driven inconsistency of policy as well.) Conversely, the Framers conferred some legislative power on the president in the form of the veto. And they mixed both presidential and congressional prerogative into the judiciary branch by making the president responsible for nominating persons to the federal judiciary, with actual appointment subject to final confirmation by the Senate.

And they concocted the contraption known as the Electoral College.

* * *

The delegate to the Constitutional Convention who advocated most strongly for a robust executive was James Wilson. He had immigrated to Pennsylvania from Scotland in the 1760s, bringing with him a deep immersion in the ideas of the so-called Scottish Enlightenment, among whose most prominent figures were David Hume and Adam Smith. Wilson and his colleague James Madison are often bracketed as the two most sophisticated political theorists at the convention. And Wilson and Alexander Hamilton were the foremost proponents of what Hamilton called "an energetic executive."

Here's what Hamilton had to say about executive energy in *Federalist* no. 70:

> A feeble executive implies a feeble execution of the government. A feeble execution is but another phrase for a bad execution: And a government ill executed, whatever it may be in theory, must be in practice a bad government.[2]

Most notably, Wilson wanted direct popular election of the president—and for good reason. In his view, the presidency was the sole locus in the entire political system where responsibility for the nation as a whole resided—as distinguished from the parochial interests of representatives from local congressional districts or senators from individual states.

Wilson later reflected that no part of Constitution-making was more perplexing than the mode of choosing the president. But his conception of the presidency—which I will call "plebiscitarian," that is, an office to which persons should be elected directly by the entire citizenry, and in which they should be directly beholden to the national at-large electorate—was not to be, at least not for roughly the first century and a half of nationhood. Instead, we got the decidedly odd and distinctly American apparatus of the Electoral College.

The Electoral College is a vestigial votive offering to federalism, an antique artifact crafted in the Philadelphia compromise-factory of constitutional drafting. It remains to this day a mystifying piece of political machinery, no less perplexing to Americans themselves than it is to foreign observers (just try explaining it to an inquisitive foreign visitor). And it is but one of several reminders of how the Founders both longed for and feared meaningful executive power—and of how they regarded the voice of the people with both reverence and dread. But—just to look ahead for a moment—James Wilson's aspiration for a more plebiscitarian presidency would in the fullness of time get a kind of second wind.

But back to the eighteenth century, and to the Constitution crafted at Independence Hall. Some further numbers can serve to make a cardinal point. Article I of the Constitution addresses the role of the legislative branch. It comprises fifty-one paragraphs. No less significantly, it contains language about "Powers Denied to the Government," suggesting an elision in the Founders' minds between "government" and "legislature."

Article II addresses the executive branch. It contains just thirteen paragraphs, eight of which lay out the mechanisms for electing the president and four of which detail his "powers." One provides for his impeachment.

The asymmetry of those numbers—fifty-one paragraphs devoted to the legislature and just thirteen to the executive—strongly suggests that the Framers conceived of the president as largely the creature of the legislature. He was to be a political agent with some autonomy, to be sure, and he was to inhabit something quite different from a parliamentary system, where the head of the majority party in the legislature is also the head of government. But he was intended to be an actor who, though independently elected, would in practice be substantially subordinated to the will of the legislative branch.

That interpretation of the Founders' intent is reinforced when we recollect that down to the 1830s presidential candidates were chosen by congressional caucuses. Following the disappearance and presumed murder in upstate New York of William Morgan, an outspoken opponent of Freemasonry, the newly formed and short-lived Anti-Masonic Party convened what is generally regarded as the first presidential nominating convention in Baltimore in 1832. It proposed a distinguished Virginian and former US attorney general, William Wirt, for the presidency. Other parties soon followed suit. Nominating conventions, composed of delegates drawn not just from the membership of Congress but from the broader electorate, signaled the rise of mass democracy in the Jacksonian era. They heralded a growing demand for more popular access to the political system in general and to the presidential nominating process in particular. The fact that the first party to hold a convention was animated by deep suspicion of the presumably elitist and secretive Freemasons provides an early clue as to the strength of "populist" sentiment in American political culture. It is no coincidence that the president most conspicuously associated with the populist strain in American politics was Andrew Jackson, the victor over both Wirt and Henry Clay, the "National Republican" candidate, in 1832.

The Jacksonian era is rightly regarded as the *fons et origo* of several popularizing strains in American political culture that have persisted and, indeed, amplified over the course of American history—from the preference for leadership molded from common clay and the concomitant suspicion of elites to the privileging of local over central power. Those nominating conventions born in the Jacksonian era also had a long, though ultimately limited, life span. They continued to play a role in choosing presidential candidates for just over a century after 1832. The last convention that went to a second ballot was the Democratic convention that nominated Adlai Stevenson in 1952. Since then, and especially since 1968, Americans have dwelled in a different political landscape, where the conventions have become largely superfluous or redundant infomercials with no real decision-making consequence.

So the presidential system born in the eighteenth century persisted more or less intact well into the subsequent century. It may well be the case that the average college student of American history today could name more prominent congressional figures than presidents from the nineteenth century. Yes, many could cite Jefferson, Jackson, and Lincoln—but Millard Fillmore, Franklin Pierce, James Buchanan, Chester Arthur, Benjamin Harrison, and many others are not names that reside prominently in core memory. But I'll bet that lots of students would recognize the names of nineteenth-century congressional grandees like Daniel Webster, John C. Calhoun, Henry Clay, Stephen A. Douglas, Charles Sumner, Thaddeus Stevens, or Thomas B. "Czar" Reed.

In any case, for better or worse, Congress remained both in fact and in popular perception the seat and solar plexus of American governance at the federal level well into the nineteenth century.

As John Hamre has observed, our problems and challenges, both domestically and globally, are mainly horizontal, crossing all sort of lines. But our organizations, our governance is vertical.

—Sam Nunn

But toward the end of that century, several observers began to wonder if the legacy constitutional relation between Congress and president was not obsolescing. As America's industrial revolution gathered phenomenal momentum and commerce and communication expanded to continental scale, as people moved in droves from countryside to densely packed cities, as immigrants came ashore in ever-larger waves, the felt need for a Hamiltonian "energetic executive" grew increasingly acute. Among the earliest proponents of radically rethinking the American governmental system was a young graduate student at Johns Hopkins University. In 1885 he published his doctoral dissertation under the title "Congressional Government." It remains to this day one of the most trenchant treatises ever written about American political institutions.

That bright young graduate student was Thomas Woodrow Wilson. He was, of course, destined some three decades later to become the twenty-eighth president of the United States.

Wilson intended his title, "Congressional Government," to be understood as ironic, even oxymoronic. His central argument was that Congress was inherently—structurally—incapable of anything resembling coherent, effective government. As he wrote of the US Congress:

> Nobody stands sponsor for the policy of the government. A dozen men originate it; a dozen compromises twist and alter it; a dozen offices whose names are scarcely known out of Washington put it into execution . . . [yielding] the extraordinary fact that the utterances of the Press have greater weight and are accorded greater credit, though the Press speaks entirely without authority, than the utterances of Congress, though Congress possesses all authority. . . . Policy cannot be either prompt or straightforward when it must serve many masters. It must either equivocate, or hesitate, or fail altogether. [The] division of authority and concealment of responsibility are calculated to subject the government to a very distressing paralysis.[3]

You have to pinch yourself to remember that those words were written not in 2017 but 132 years ago, in 1885.

Wilson's voice was an early one in a chorus of similar commentary over the next few decades, culminating in works like Herbert Croly's *The Promise of American Life* (1909), which notably advocated "Hamiltonian means to Jeffersonian ends," and Walter Lippmann's *Drift and Mastery* (1914), which is subtitled, tellingly enough, "An Attempt to Diagnose the Current Unrest." Like Wilson, Croly, Lippmann, and many others in that so-called Progressive Era had come to believe that Congress was by its very nature too fragmented, too unaccountable, its powers too dispersed and occluded from public view, its constitutional character too deeply rooted in localism and parochial interests, to be capable of coherent governance on a national scale. All that might have been tolerable in the republic's youth, Croly and Lippmann argued. They noted that the Constitution was drafted when the United States was a nation of farmers, and when but four million people inhabited a territory about one-tenth the size of the twentieth-century United States. But the chronic disarticulation and lamentable dysfunction of Congress now that the United States was a big, mature, complex, urbanized, increasingly networked and interdependent industrial society of nearly one hundred million increasingly diverse people, with the capacity to assert its influence on a global scale, was both an embarrassment and a danger.

But the presidency—there, thought Wilson and Croly and Lippmann and Theodore Roosevelt and many others of their generation—the presidency was the fulcrum that, if managed with creativity, muscle, and art, could be made to serve the larger interests of this big, continentally scaled, ambitious, energetic society.

So when he assumed the presidency in 1913, Wilson represented both a newly emerging popular conception of the president's role and a new style of presidential leadership. Theodore Roosevelt had prefigured these changes, but Wilson significantly consolidated and advanced them. As he said, "The President is at liberty, in law and in conscience, to be as big a man as he possibly can."[4]

That statement was more aspirational than descriptive in Wilson's day, but it nonetheless set the compass headings for almost all future presidents. Between them, Theodore Roosevelt and Woodrow Wilson introduced two significant innovations to the presidency.

The first is evident in the fact that with Roosevelt we have the first publicized slogan—the Square Deal—that described a comprehensive, coherent policy program for which the president was to stand as champion. No such thing existed before the twentieth century. But Americans have long since become accustomed to—indeed, have come to expect—presidentially sponsored policy packages, along with their headline slogans—from Theodore Roosevelt's Square Deal, to Woodrow Wilson's New Freedom, FDR's New Deal, Harry Truman's Fair Deal, JFK's New Frontier, Lyndon Johnson's Great Society, and Donald Trump's Make America Great Again.

That succession of presidential programs bespeaks the felt need in modern American society for the type of coordinated, articulated, national policies for which the lone national political officer can be held accountable. But of course, while the president proposes, Congress disposes, and it has many, many avenues of disposal. Here is where constitutional realities and the abundant liabilities of what Francis Fukuyama calls the American "vetocracy" continue to come frustratingly into play. As the young Woodrow Wilson observed, Congress to this day remains the place where presidential policy initiatives go to die—or, perhaps even more regrettably, to be disemboweled or dismembered beyond all recognition.

The second innovation whose outlines, at least, we can see in the Progressive Era of Roosevelt and Wilson recalls that plebiscitarian dream of James Wilson back in 1787. The Electoral College, and the full panoply of congressional methods and mystifications, of course, continued to abide. But both Roosevelt and Wilson began to develop a political technique that would grow exponentially in incidence and effect as the twentieth century went forward: using *publicity* as a tool of governance. And here is where the history of the presidency and the history of communications technologies fatefully intersect.

The problem for the modern decision-maker is, he has to decide, to pick one of these courses of action, given a wicked problem, fully aware he might be totally wrong.   —T. X. Hammes

---

By publicity, I mean reaching over and beyond Congress to appeal directly to the public at large in order to advance the presidential agenda.

The emergence of inexpensive mass-circulation newspapers around the turn of the century—papers like William Randolph Hearst's *New York Journal* and Joseph Pulitzer's *New York World*—first made this possible. Wilson conspicuously used a kind of saturation publicity to mobilize public opinion and compel passage of tariff, banking, trade, and antitrust legislation in his first term. He tragically broke his health in a failed attempt to do the same with respect to ratification of the Treaty of Versailles in 1919.

That shift in the political uses of communication technologies has been precisely quantified. Research confirms that in the twentieth century, presidents spoke directly to the public—in one medium or another—six times more frequently than in the nineteenth century. Conversely, presidents in the twentieth century spoke exclusively to Congress one-fourth less frequently than in the preceding century.[5]

The later emergence of mass electronic, instantaneous communication—that is, the radio—powerfully accelerated that trend. Franklin D. Roosevelt, of course, with his renowned Fireside Chats, fundamentally redefined the president's relationship to the public. He deliberately relied on the radio to end-run the newspaper magnates like Hearst, whom he considered his political adversaries, and to mobilize public opinion to bend Congress to his will.

Roosevelt's use of the radio to speak directly to his fellow citizens in real time represents one step in a continuing process of political "disintermediation" (or the removal of intermediaries who once delivered, interpreted, or commented on communications between leaders and citizens) that in our own time has fantastically accelerated. Citizens today increasingly receive their political news directly from the politician's mouth, Twitter finger, or Facebook page, without filtration by editors or reporters. In Roosevelt's case, radio displaced the earlier mass-communication technology of high-circulation newspapers that had served Wilson so well. In time, John F. Kennedy advanced this process still further when he began televising news conferences, rendering the evening broadcast or the next morning's print account of presidential pronouncements utterly redundant. (Though, ironically, the presidential preemption of "news" exponentially expanded the volume of "commentary" to fill traditional news holes.)

The internet and social media of our own time take this process of disintermediation to an extreme conclusion. They not only provide presidents (and presidential candidates) with direct access to citizens, but also enable citizens to communicate directly and swiftly with leaders—and, even more consequentially, with one another, free from editorial curating or fact-checking or even the protocols of civil speech.

Contrary to past predictions that the proliferation of mass media would nurture a "global village," in fact the multiplicity of media is dividing us into ever more isolated tribal units—and is resurrecting primal habits of trusting no one outside the tightly straitened circles of the familiar. Distrusting government—indeed, distrusting all institutions and persons beyond one's own immediate orbit—is an old American habit. As Edmund Burke said of the rebellious Americans in 1775, "The religion most prevalent in our northern colonies is a refinement on the principle of resistance; it is the dissidence of dissent, and the Protestantism of the Protestant religion."[6] But even Burke would be astonished at how deeply dissidence and distrust have taken root in twenty-first-century America. Americans today not only distrust governmental as well as a broad array of other institutions—but they increasingly distrust one another. Recent polling data confirm that distrust is pervasive in our society and, alarmingly, that young people are the least likely to place trust in others.

Consider the following data (table 8.1 and figures 8.1–8.2), especially unsettling in light of Alexis de Tocqueville's warning nearly two centuries ago that "despotism . . . is never more secure of continuance than when it can keep men asunder; and all its influence is commonly exerted for that purpose. . . . A despot easily forgives his subjects for not loving him, provided they do not love each other."[7]

TABLE 8.1   Confidence in US Institutions, 2015, vs. Historical Average Since 1973 (1993 in some cases)

Percent of American public expressing "a great deal" or "quite a lot" of confidence in each institution.

	Historical Average		2015
Military	68%	↗	72%
Small Business	63%	↗	67%
Police	57%	↘	52%
Church / Organized Religion	55%	↘	42%
Medical System	38%	↘	37%
Presidency	43%	↘	33%
US Supreme Court	44%	↘	32%
Public Schools	40%	↘	31%
Banks	40%	↘	28%
Newspapers	32%	↘	24%
Organized Labor	26%	↘	24%
Criminal Justice System	24%	↘	23%
TV News	30%	↘	21%
Big Business	24%	↘	21%
Congress	24%	↘	8%

Source: Jeffrey M. Jones, "Confidence in U.S. Institutions Still Below Historical Norms," Gallup, June 15, 2015, http://news.gallup.com/poll/183593/confidence-institutions-below -historical-norms.aspx.

FIGURE 8.1   The Decline of Trust in the United States

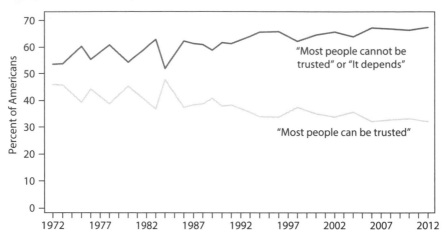

Source: Josh Morgan, *Medium,* 2014, from General Social Survey data, accessed December 7, 2017, https://medium.com/@monarchjogs/the-decline-of-trust-in-the-united-states-fb8ab719b82a.

FIGURE 8.2   The Decline of Trust among the American Public by Generation, 1972–2012

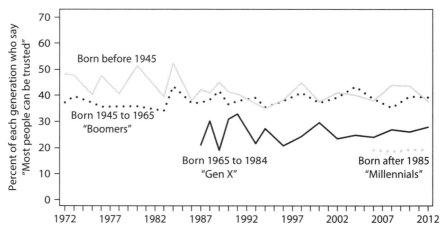

Source: Josh Morgan, *Medium,* 2014, from General Social Survey data, accessed December 7, 2017, https://medium.com/@monarchjogs/the-decline-of-trust-in-the-united-states-fb8ab719b82a.

When combined with the hyper-fragmentation of media outlets that the internet has facilitated, social media has unleashed all the perils of what psychologists call "confirmation bias," or the quite natural but also troublesome human tendency to give more credence to those views that reconfirm one's already existing views and to discount deeply all contrary voices. The proliferating narrow-casting, even micro-casting, that characterizes today's supersaturated media environment lends a kind of perverse proof to the old maxim that more is not necessarily better.

There is one further development that might be subject to the "more is not necessarily better" critique. I'm referring to another innovation that dates from the early twentieth century—the proliferation of primary elections. Oregon held the first delegate-binding presidential primary election in 1910. California and a handful of other states soon followed suit. They did so in the name of "direct democracy"—taking politics out of the hands of the "bosses" and "machines" and putting power squarely into the hands of the people.

In American political culture, it's hard to argue that more democracy is not better than less democracy. But the actual workings of the primary system might prompt us to rethink that apparently benign proposition.

The fact is that, as late as 1968, only a dozen states held presidential primary elections. A decade or so later, virtually every state had a primary—or its near equivalent, a caucus.

Here is another instance of disintermediation, with far-reaching consequences. While the electronics revolution has severely reduced the influence of the established press and other media, in more or less the same time frame primary elections have enormously reduced the role of political parties in performing their usual tasks of identifying, vetting, recruiting, grooming, and supporting candidates. Now, any political entrepreneur with a fat checkbook or a few fat-cat supporters can seek to "rent" a party as the vehicle of his or her candidacy—a consideration that helps explain why the Republican field in the 2016 election cycle had seventeen contenders, many of whom had sufficient funding to hang on well beyond their sell-by date.

When people talk about this being the beginning of the "post-Western" era, you may be saying in effect the "post-moderate era"—an era beyond when we can count on political parties to bring us to a moderate position.   —Jim Hoagland

This arrangement amounts almost exactly to an inversion of the historical relation between would-be candidates and parties. Some might even argue that "rent" is too weak a word—that in today's environment it has become possible to "hijack" a party.

So here is where history has deposited us: Americans have come to have increasingly extravagant expectations not only that the president will protect them from each other's congressmen but will also be the paladin of coherent nationally scaled policies, domestic and foreign, responsive to the realities and the responsibilities of an advanced post-industrial society of 322 million people. In the absence of tempering, mediating institutions like a responsible press and functioning political parties, presidential aspirants can nourish those expectations as they will, but without meaningful appraisal, expert analysis, or restraint.

This is plebiscitarianism run amok, plebiscitarianism without the benefits, because the American political system as a whole proves stubbornly unable to satisfy those expectations. And like presidential candidates, congressional representatives are increasingly free agents, only weakly shepherded by party "leadership." Yet Congress retains all its prerogatives to obstruct and to veto. It continues to operate as a ramshackle confederation of local interests rather than a truly national legislature. The resulting stalemate feeds public frustration, disillusionment, distrust, and resentment and breeds the political attitude we call populism. It is not a pretty picture.

A recent book by Terry Moe and William Howell, *Relic: How Our Constitution Undermines Effective Government—and Why We Need a More Powerful Presidency*, resurrects much of Woodrow Wilson's lament about congressional inefficiency in 1885.[8] Moe and Howell go further than Wilson, however, and propose a quite specific remedy:

granting the president across-the-board "fast track" authority with respect to all legislation, such as he now enjoys with respect to trade negotiations. If their recommendation is adopted, presidential initiatives would have to be voted up or down, without amendments or riders. In their view, this arrangement would better align the hopes invested in the presidency and the realities of presidential leadership by meaningfully attaching presidential accountability to presidential promises— as James Wilson wanted—and would introduce more transparency to the legislative process—as Woodrow Wilson wanted—while still allowing Congress to retain its essential powers even while mitigating its capacity to obstruct.

Whether this is a realistic, or a sufficient, or even an appropriate solution to the political paralysis that besets the United States today, I can't say. But any diagnosis of the current unrest, to borrow Walter Lippmann's phrase, must take account of the mighty weight of constitutional architecture, technological change, institutional evolution, and historical practice that has brought the American political system to its present sorry pass. If some means are not found to establish reliable credibility in our organs of information and communication, revivify the capacity of political parties to channel and responsibly focus citizens' interests, restore confidence in our principal institutions and our trust in one another, and, finally—as was the goal in the Progressive Era—to align the authority of the executive and other branches of government with the realities of contemporary life, the great American political experiment, so hopefully launched more than two centuries ago, may well be doomed.

9

TECHNOLOGICAL CHANGE AND LANGUAGE

Charles Hill

anguage has been regarded as the original and most fundamental human tool. At certain points in history when other aspects of technology have enhanced or damaged the use of language, major changes in world order have resulted. We are witnessing such a phenomenon now, with impacts on individual psychology, on socio-pathologies, on autocratic regime powers, and on democratic governance.

At present, a "language revolution" is under way, propelled by an eruption of electronic communications technologies that, while enhancing productivity, are also creating social and political chaos. This phenomenon cannot be successfully understood or managed without an awareness that the modern age itself, beginning some three hundred years ago, has been defined and shaped by the tension between *thought* and *things* in a contest for control over the languages of communication. This chapter aims to describe this struggle across history as a way to shed light on the present situation.

The e-revolution in communication is now challenging, even threatening, the conduct of responsible governance: marginal sociopaths are being empowered to organize and act collectively as never before; dictatorial regimes are perfecting powerful tools to surveil and suppress entire populations; and instantaneous popular judgments on political issues are beginning to overwhelm representative government as designed by the Founders to avoid the chaos-producing "direct" democracy feared in premodern societies.

> *"It's just another language."*—An alumnus who majored in humanities when asked at his tenth class reunion how he could be so successful in computer science "coding," never having studied it.

> *"There can be no liberty for a country which lacks the ability to detect lies."*—Walter Lippmann

Aristotle declared that "man is the political animal"; Adam Smith countered with "the trading animal." Across the centuries, however, most thinkers have regarded man to be "the language animal."

The stereotypes of national languages have a certain truth to them. Latin is compact. German is convoluted. Italian is exuberant. French is precise. Japanese and Hebrew rely on a distinctive part of the brain. Chinese tones and ideograms are distinctively difficult to master.

A language can collapse. Thucydides's *Peloponnesian War* is an extended analysis of how misused language deteriorated until it led to the destruction of a great democratic empire. A language can also be driven out of control, as revealed by Tacitus's *Annals* of Nero's Rome or George Orwell's "Newspeak" in the novel *Nineteen Eighty-Four*. At turning points in history, language has required "rectification" (Confucius). Words have been deliberately reversed in their definition (Machiavelli) or used as weapons (Robespierre).

Language produces narratives that form the basis for cultures, which do not remain static across time but create and operate within paradigms that, at certain revolutionary periods, have dramatically shifted to bring fundamental change to the human condition.[1] Many of the greatest linguistic and technological revolutions have occurred in lockstep:

- *The Agricultural Revolution* revealed that the earth could be torn up—plowed and technologically induced—to produce crops, yet also illustrating that the ground beneath was no longer sacred or "enchanted."

- *The Scientific Revolution* came when new instruments meant that the human body itself could be torn up—vivisection—to be studied through a new language of measurement based not on awe and wonder but on observable and knowable units.
- *The Industrial Revolution* ran on two engines: physical energy sources that could be newly discovered and harnessed and a language of production and mechanics that could serve large populations.

The other revolutionary category, joining those of science and technology, is humanistic—changes in human consciousness that, just as with the other revolutions, have required language as their chief vehicle. This was stated most influentially by the German philosopher Hegel. Premodern consciousness had been shaped by theology. For the modern world, *theology* would be replaced by *history* as the arena in which the most profound human problems would have to be faced. Hegel declared, "History is the history of humanity's increasing consciousness of freedom."

- *The Renaissance*, conventionally set around 1500, is understood to mark the shift from the proper study of mankind being the City of God to study of the City of Man.
- *The Reformation*, the 500th anniversary of which was in 2017, added individual decision-making in a vast new public and private arena opened by the ending of the church's temporal jurisdiction.[2]
- *The Enlightenment*, as declared by Kant in 1784, ended all Foundations, that is, all beliefs, traditions, texts, and hierarchies that peoples in the past depended upon for answers to life's questions. From that point forward, only Reason would be intellectually legitimate; external sources of authority would be disparaged.

Mark Zuckerberg's vision of a global community that will be created when we're all connected is incredibly similar to Martin Luther's notion that there could be a priesthood of all believers.

—Niall Ferguson

Making a "Modern" Age

The tensions and connections between these two revolutionary categories—technologies and humanities—would be negotiated by language and fought over for control of language.

Taken together, these two revolutionary chains made it clear that a new "modern" era would have to be invented. This would in the first instance be done by "the book," which, in its codex form, would become the technology best suited for presenting "an extended argument" for what would supersede the ancient and medieval conception of life.

The technologies of movable print initiated by Gutenberg would be massively disruptive by empowering the book as a mechanism for instruction. This is dramatized in Victor Hugo's seminal novel, *Notre-Dame de Paris*, known to English readers as *The Hunchback of Notre Dame*. The cathedral is the teacher of the masses, the illiterate peasantry. Quasimodo, the deaf and near-blind sacristan, knows every statue, shrine, gargoyle, and saintly bas-relief of the edifice as he scrambles all over it, not hearing but feeling the power of its mighty bells. To him, the cathedral is a living organism, a living language. But the book will bring an end to the Age of the Cathedral for, as Frollo says, holding a book and viewing the church, "*This* will take the place of *that*."

As the book brought with it the Enlightenment, the times being new, it was imperative to think anew.[3] This raised the question of what meaning, what new definition, we would give to what we see around us. This would require the closest observation of nature, drawing and adapting inferences from it. With what previously was thought to *be* natural now declared *not to be* natural, humanity seemed to require the invention of a new entity: "the artificial." For example, "the divine right of kings," a premodern belief as a natural, God-given power, was no longer acceptable. Modern political theory thus had to create an "artificial person" to fulfill the role of sovereign power over the modern state. The great works of the Enlightenment, each set forth in the ever more important new technology of the book, reveal their authors' struggles with the tensions between "the natural" and "the artificial" as a problem of language.

Seven very big books of the Enlightenment—all in English for the sake of intertextual coherence—reveal a determination to remake the meaning of every field of study, both humanistic and scientific.

1. Thomas Hobbes's *Leviathan*, 1651, was written to provide a theory for the new basic fact of geopolitics, the state (declared by the Treaty of Westphalia in 1648 to be the successor to the empire). Hobbes starts by revising human nature itself. *Leviathan*'s very first sentence declares that man—putting aside Aristotle's claim that "man is the political animal"—can make an "artificial" animal: in other words, an artificial human being. To replace the divine-right king (Charles I had just been beheaded by the English Revolution), an artificial sovereign was "created" to do the job. Individual human kings may come and go, but the artificial sovereign would remain as the highest, perpetual, indivisible political power.

2. William Blackstone's *Commentaries on the Laws of England*, 1765–1769, replaced the organic, naturally exfoliating, centuries-old English common or case law with a fixed compendium of statute-like assertions. The old English Common Law had grown organically, as if it were divinely inspired, part of God's Plan. Blackstone's tome in four volumes was unequivocally man-made law. The concept of the "artificial" appears early on in Blackstone's work and would have far-reaching influence in the United States, ensuring that the Common Law would not take root here.

3. Adam Smith's *The Wealth of Nations*, 1776, preceded by his less famous but equally profound book *The Theory of Moral Sentiments*, 1759, relies almost entirely on two artificial and, indeed, imaginary

conceptions: "the invisible hand" would turn every individual's self-interested economic activity into wealth for the community as a whole and "the impartial spectator" would invisibly observe you so that your awareness of being scrutinized would incline you toward ethical behavior. Together, these concepts would turn traditional ideas of wealth creation (the "mercantilism" Smith was determined to overthrow) upside down. This would be called "asocial sociability,"[4] the claimed discovery that when every individual acts materially in his own particular interest, the benefits to the commonwealth as a collectivity will be enhanced. This would rightfully become economic orthodoxy for much of the modern world.

4. Samuel Johnson's *A Dictionary of the English Language*, 1759, a monumental endeavor requiring immense personal and intellectual labor, fixed the meaning of words and, beyond that, the language itself. The book also would serve as a form of encyclopedia by using words as anchors for cargo ships of knowledge. No longer would spellings and meanings be permitted to vary with each individual speaker or writer; a "standard" language was in the making, to be carefully watched over by designated specialists. When paired with Blackstone's *Laws*, Johnson's magnum opus would embody the Enlightenment mind, taking an immense realm of natural, organic evolution—language and law—and confining it in a supposedly rational form as an "artificial" foundation.

No mention of Samuel Johnson can avoid reference to James Boswell's massive 1794 *Life of Johnson*, the greatest biography in literary history and plausibly the work that gathers in one telling the most we can know about the life of any human being. It is as though Boswell, aware of the felt need to create "artificial" persons, had decided to make exhaustively clear what a truly natural person would be like—Samuel Johnson.

5. Edward Gibbon's *The Decline and Fall of the Roman Empire*, 1776, signaled the Enlightenment's replacement of theology by history as the arena in which humanity's greatest and most consequential questions would have to be addressed. Gibbon used his history to clear away the flaws, follies, and foundations of the past in order to provide the mod-

ern era then in the making with a clear field for new—artificial—departures. The zeal and institutions of Christianity, that great foundation, had weakened even as it aggrandized the Roman Empire. But Rome's artistic and humanistic influence lingered on and prompted the supreme importance of order over the chaos of a thousand years. As an exemplar of the Enlightenment, what was most important to Gibbon were books. A book was the work of a man, and Gibbon's respect and reverence were unbounded. *The Decline and Fall* is a bibliographical survey of European civilization from the second to the fifteenth century. All the great writers are commemorated, all the famous books are noticed, and on each occasion, Gibbon carefully tells his reader what his own reactions are to these creations of man.

The Life of the Mind is the hero of the *Decline and Fall*, the mind enshrined in a book, the exemplar of which is *the book, Decline and Fall* itself.[5]

At the same time, Gibbon closely observed the practices and skills of those who make the world work. As he famously said of himself, "The discipline and evolutions of a modern battalion gave me a clearer notion of the phalanx and the legion; and the Captain of the Hampshire Grenadiers (the reader may smile) has not been useless to the historian of the Roman empire."[6]

These books of the Enlightenment possess a striking intertextuality or cross-referencing of interests. Adam Smith regarded the great reconnaissance of the world in the early modern period as providing unparalleled significance for humanity:

The discovery of America, and that of a passage to the East Indies by the Cape of Good Hope, are the two greatest and most important events recorded in the history of mankind.[7]

This was because, Smith said, the communication and commerce of the species as a whole enabled all mankind "to relieve one another's wants, to increase one another's enjoyments, and to encourage one another's industry."[8]

6. In this context, Captain James Cook's *Journals*, 1769–1979, is worthy of joining this pantheon of big books of the Enlightenment because it bears out Aristotle's attention to the "practical arts" as vital sources of knowledge; it is a matter, as John Dewey would much later conclude, of "learning by doing."

Aristotle declared that statecraft is a practical art much like navigation. Cook navigated, learned from it, and wrote incessantly and at length about it all. Dispatched by the British Admiralty to Tahiti to observe the transit of Venus, he eventually used the much-perfected new chronometer to fix the precise location of Tahiti, a matter of great importance to political and geostrategic comprehension, for it had been believed that islands had no constant location on the seas. In fixing Tahiti's position and mapping its coastline boundaries, Cook contributed to the modern concept of a state as a single territory with recognized borders.

Cook recorded in careful detail the physical world of lands and seas in his three voyages to the Pacific and Southern Oceans, surveying and charting New Zealand, claiming Australia, assaying the extent of Antarctica, and seeking a northwest passage across the Western Hemisphere. Cook's explorations created a breakthrough in instrumentation. He enabled the measuring of what could not be measured before and the raising of the consciousness of the world's peoples about the planet they inhabited.

7. Charles Darwin published *The Voyage of the Beagle* in 1839 and *On the Origin of Species by Means of Natural Selection* in 1859. As Cook minutely observed the geographic physical world, so Darwin, the ultimate "naturalist," would observe the doings of living creatures, however minute they might be, and brought the Enlightenment revolution in language to the realm of natural science. As Darwin wrote in his *Autobiography*:

The Voyage of the Beagle has been by far the most important event in my life and has determined my whole career. . . . I have always felt that I owe to the voyage the first real training or education of my mind. I was led to attend closely to several branches of natural history and thus my powers of observation were improved, though they were already fairly developed.

No one had ever observed "nature" as completely and in such detail as Darwin. In doing so, he added a dimension to human consciousness.

Such voluminous factual observations of nature called for an explanatory theory, an "artificial" conclusion that could be tested against further scientific studies. Darwin would argue that species evolve, or change. This would be published twenty years after the Beagle's voyage as The Origin of Species. Here, science would predominate over theology and swiftly provide an entirely new framework, that of reason, about the creation of the world. The "natural" could only be understood through the description of an "artificial" encompassing concept. This could only be done by words, by language.

In New England, Emerson and Thoreau would see this differently. Emerson's greatest work, "Nature," would not be the product of a "naturalist," as was Darwin, but of a Transcendentalist. Was the result artificial? Or was it a natural result of the mind conducting itself naturally? Thoreau would, unknowingly, provide an alternative to Cook's and Darwin's measurements. Thoreau would measure Walden Pond just as carefully as they would have done, yet in a way that would lead to the conclusion that you cannot "know" the pond by measuring its depth nor "know" a woodchuck by cooking and eating it.

Emerson and Thoreau were doing what Wittgenstein much later would do: demonstrate the importance of "elimination work"—that is, to show the limits of what scientific observations and subsequent reasoning can do in producing the knowledge most worth knowing.

It is interesting to contemplate the different ways that, as contemporaries, two "naturalists," Darwin and Emerson, observed nature. Darwin saw "species" as nature's classifications and recognized that species change. Emerson saw nature as providing the soul with a ladder up which to ascend, from nature as a commodity, to serve our lowest needs, to beauty, which is an object of the intellect, which uses language to conduct a *discipline* striving for *idealism*; the final synthesis is *spirit*.[9] Significantly, Emerson's "specular mount" up which the soul ascends is centrally dependent on language—number four of the seven steps—as the technological tool required in the process.[10]

Emerson saw a kindred soul in Goethe, as "the writer" extolled in Emerson's essays on *Representative Men* summed up in one sentence: "Goethe would have no word that does not cover a thing."

Goethe's great work *Faust* would expound upon this foundational problem of the human condition: how language succeeds or fails to derive or infuse meaning from or for the things and acts of this world. Early in *Faust* I, Doktor Faust, surrounded by books in his study, rejects them and all learning, declaring, "In the beginning was *not* the Word!" and concluding that instead it was "The Act!" (*die Tat!*). Here was a call for revolution in action. As Marx would say, "The point is not to understand the world but to change it."

At the end of *Faust* II, however, Goethe wrote four mysterious yet profound lines that have baffled translators' efforts to convey the imperative to interrelate words and things for the production of meaning:

Das unbeschreibliche
Hier ist's getan;
Das ewig-weibliche
Zieht uns hinan.

Not a literal translation, but meaning, as Emerson would sense:

Here deeds understand
Words they are shadowed by.[11]

So, in reviewing the modern project, are actions rightly represented by their labels? Do words accurately describe the deeds they cover? If the technologies and the humanities are linked by language, are they also engaged in a rivalry over the command of that vocabulary?

Should we be stopping scientific experiments? Should we have stopped Galileo? Should we have stopped Copernicus?
—Lucy Shapiro

Testing the Modern Artificial

Somehow—it will forever remain a mystery—the many conceptions and considerations of the time were sensed and made into a novel by an eighteen-year-old girl named Mary Wollstonecraft Godwin, soon to take the name of her husband, the Romantic poet, and be known as Mary Shelley. Done as a story-writing game to while away the time when lodging on the shores of Lake Geneva on chilly summer evenings, it would become world famous as *Frankenstein*, published in London in 1818.

Strangely, to preface this uniquely odd contribution to literary and cultural history, the book was given a short preface by Percy Bysshe Shelley that began by noting that "Dr. Darwin" (Charles Darwin's grandfather and precursor in evolutionary theory) believed that the tale of Frankenstein's monster was "not of impossible occurrence."[12]

The novel would become a never-ignored myth referred to across the years in other novels—the "Frankenstein" concept shimmers through Dickens's *Great Expectations*—and plays, films, cartoons, advertisements, and virtually every form of visual and written expression; it would become ensconced as a standard text in the canon of literature and intellectual history.[13]

In fact, Mary Shelley's *Frankenstein* is the key text for understanding the Modern Age. In it, "the artificial" is created as the product of technology and humanity and then displays the inherent tension between the two. Contrary to the accepted interpretations, this conveys far more than the tale of a scientist who concocts a monster that cannot be controlled and destructively roams the earth. Yes, there is something at that level, but the novel's intricacies are far more complex than that, as it, in

Lawrence Lipking's words, "furnishes a testing ground for every conceivable mode of interpretation."[14]

The most consequential interpretation can be stated: the Monster,
once in being, possesses the vocabulary and the emotion of a human
search for understanding and affection, yet is demonized and rejected by
humans themselves. The artificial proves more human than the human.

Language is at the center of the encounter between the human and the
scientific. Frankenstein the scientist used human body parts to create an
ugly being who nevertheless feels and seeks human love. As observed
by Peter Brooks, the Monster grasps the nature of language as a system,
"as both the tool he needs to enter into relation with others, and a
model of relation itself . . . that from which he feels himself excluded."
The Monster tells Frankenstein:

> Although I eagerly longed to discover myself to the cottagers, I
> ought not to make the attempt until I had first become master of
> their language which knowledge might enable me to make them
> overlook the deformity of my Figure.[15]

But his innate capacity to love is thwarted at every turn by his creator,
who made him ugly, and by the culture of scientific modernity.

---

*I think in our heart we must allow people to accept more risk on
developing new technologies—and, if they fail, to get back on the
horse and keep trying. I do believe that we can do this, and I do
believe we once embraced that capability, but we have let it slip
away.*   —James O. Ellis, Jr.

Books and Their Arguments Undermined, Transformed

A second look at the Big Books of the Age of Enlightenment reveals
what might be called "the struggle of—and for—the modern age." A

twenty-first-century assessment of these works indicates that the project of the modern is troubled or is being transcended.

Law

The legal system is now statutory, with an increasing proliferation of regulations and major legislation voluminously set forth to far exceed Blackstone's *Commentaries*, which summed up centuries of common law.

The US Constitution, as an "artificial foundation" produced in the American Enlightenment, has become an interpretive battlefield on which two different "languages" vie for Supreme Court power. One holds words as written to be determinative, whereas the other holds words to mean what we want them to mean at the present moment (as Humpty Dumpty said, "When I use a word, it means just what I choose it to mean—neither more nor less").

Economy

The major phenomenon of the late twentieth and early twenty-first century has been globalization, which may be seen as Adam Smith's division of labor carried to a planetary scale. Smith's reasoning was production-based, whereas current signs suggest massive shifts to consumption-based societies. Smith's primary, and revolutionary, insight that self-interested economic activity by individuals would benefit the economy as a whole has largely been swept away by central state regulation. Economic language has shifted from that of a supply of goods creating a demand to that of demand dictating the goods supplied. People care more about consuming than producing, encouraged by central-state policy. The Big Books on economy in this century have raised alarms about the end of growth or the inevitability of inequality even as they seem less persuasive than appeared at first reading. And what of Smith's perception about work as an indispensable contribution to morality? Can this logic hold up if a national economy's consumption is stimulated even as productive endeavor continues to decline?

Human Nature

Hobbes's Artificial Person was created to replace the divinely authorized hierarchical sovereign that the modern age would expel. Boswell's *Life of Samuel Johnson* displayed a "real" person in a full career of immense achievement despite great physical and social adversity. Hannah Arendt in the twentieth century and an array of social philosophers in the twenty-first century have, respectively, condemned or promoted the idea—traceable to Jean-Jacques Rousseau and later Marx—that human nature must be "perfected," that is, altered, if humanity is to establish an ideal polity. Modern history has revealed the imperative of "perfectibility" to be the motivating force behind totalitarianism. At present, pressures such as "political correctness" reveal that the goal of perfectibility remains significant in the minds of many. Computer science languages are enabling unprecedentedly multipliable attempts to cover all possible contingencies of life, raising the hope of "perfecting" social policies. Aristotle warned against this when considering attempts in ancient times to enact laws that would cover all future possibilities—an impossibility, Aristotle said. The computer-driven pursuit of this chimera can easily be transposed into forms of benevolent tyranny designed to make citizens conform to preordained contingent outcomes, or simply to rule out the permissibility of unforeseen contingencies.

Maybe the surprise in the AI domain is not that computers are good at making decisions and doing things—that's been true for a long time—but rather that this capability is being brought to bear on problems that I think many people thought were a long way in the future.   —Raymond Jeanloz

---

History

Gibbon's *Decline and Fall of the Roman Empire* epitomized the Enlightenment's perception that theology would, from that point forward, be relegated to the sidelines and that history would become the arena in which the greatest issues would be faced. Gibbon's volumes then interpreted the Roman Empire's final centuries during which the baleful influence of religion determined the empire's fate.

Today, however, it can be recognized that religion never went away but accompanied, permeated, influenced, and, in crucial instances, opposed the projects and purposes of the modern age. At the same time, the inclination of the secular authorities of the international state system to believe that religion had faded away or been neutered did much to delude or even incapacitate participants in governance from dealing effectively with a significant dimension of world affairs.[16] The modern inclination to disregard the claims of religion has been interpreted as depriving modern life of any legitimate meaning other than the amelioration of discomforts.

Beyond this, the modern concept of history itself has become contested as a Western-imperial concoction, denounced as a politically driven attempt to impose a "universal" or "world-historical" doctrine on the diversity of the planet's peoples.[17]

I agree with you secular is winning, but you just cannot write off religion, because on the walls of caves, people will always make some sort of divine picture.   —Bishop William Swing

---

The Practical Arts

Aristotle recognized that human beings are in a process of development driven by the necessity of their existence within the difficult and harsh demands of nature. These are the fundamental needs for protecting, acquiring, educating, increasing, and governing ourselves as people and peoples. To accomplish these tasks, people must engage in "the practical arts" such as medicine, navigation and seamanship, and agriculture and animal husbandry, as well as physical fitness through such regimens as gymnastics.

The centrality, indeed, the importance of "the artificial" is not a uniquely modern matter. Aristotle takes it up in his *Politics*, as when an individual person can become a "citizen," the latter being an artificial concept; one could be a bad person but a good citizen, and vice versa.

Underlying all this is Aristotle's contention that everything in life is in motion, ideally moving where nature intends it to go. This often was illustrated by the example of shoes—a topic of philosophical interest in ancient Athens. Bare feet obviously are natural. Does this mean that wearing shoes is unnatural? Aristotle suggests not; it is a matter of moving from lower-case natural to capital "N" Natural; in other words, an *artificial* but nonetheless legitimate *Natural*. The same was considered to hold true in the movement from subsistence farmstead to the *polis*, or city.

The test is whether the "artificial" dimension produced by the modern world is understood as the successor and substitute for the religio-philosophical "foundations" of the premodern era and whether they are positive or negative in their movement over time. But much of humanity has been out of touch with the practical arts because of technological advances that obviate the need for individuals to practice, learn from, and develop themselves this way. An example came when the creation of the global positioning system led the US Naval Academy to abandon its requirement for midshipmen to learn how to use a sextant but, after reconsideration, to reinstate that skill.

The purpose behind following Captain Cook's *Journals*, above, with Charles Darwin's *Voyage of the Beagle* was to link detailed observation of the world with hypothesis-forming to create a scientifically testable theory. Darwin's minute observations of flora and fauna, notably on the coasts and islands around South America, would eventually emerge as *The Origin of Species*, a big book swiftly recognized as "one of the major books of Western civilization," as noted by George Levine in his introduction to it. As much as anyone in the modern era, he changed human thought, his influence felt in virtually all aspects of life.[18]

Darwin's theory so thoroughly and relentlessly undermined religious convictions of millennia that his work would be vilified for generations even as it became scientifically accepted as undeniable. In this new twenty-first-century context, however, highly reputable thinkers have raised challenging objections to the neo-Darwinian account of the origin and evolution of life. According to this revolutionist approach, the process of natural selection cannot account for the actual history without an adequate supply of viable mutations, and it is doubtful whether this could have been provided in geological time merely as a result of chemical accident without the operation of some other factor determining and restricting the forms of genetic variation. As argued by the philosopher Thomas Nagel, the coming into existence of the genetic code—an arbitrary mapping of the nucleotide sequences into amino acids—with mechanisms that can read the code and carry out its instructions is improbable given physical laws alone.[19]

World Order

The modern international (Westphalian) state system can be understood as an artificial concept created out of necessity. All premodern systems purporting to establish "world order" had been substantive, imposing the rule of a particular imperial power over its region. The modern state system, however, in order to be truly accessible on a universal basis, would have to be procedural—that is, any state that agreed to adhere to a few simple procedures would be considered a legitimate international citizen and able to follow whatever substantive form of governance it chose. To make this system universal, another artificial doctrine was agreed on: the doctrine of equality of states. Obviously, no two states are ever actually equal, but the concept of juridical—that is, *artificial*—equality would be an imperative. This "artificial" can be traced to yet another Big Book, *On the Law of War and Peace* (*De Jure Belli ac Pacis*) by the Dutch jurist and diplomat Hugo de Groot, called "Grotius." Later known as "The Father of International Law," Grotius provided a structure that, emerging from the 1648 Treaty of Westphalia, would become the procedurally based international state system adopted on every continent and that is still operative, but is now in a severely deteriorated condition.

The current condition of the Westphalian international state system, the structure on which the modern world order has been maintained, has deteriorated to the point of perilousness. The question is whether it, and the modern age, can survive.

A list of artificial concepts created to replace assumptions that a disregard of theology had ruled out would include:

- Artificial language (Hobbes)[20]
- Artificial laws (from Blackstone to Oliver Wendell Holmes, Jr.)[21]
- Artificial values (as Hume concluded)
- Artificial political structure (the Constitution)
- Artificial religions (e.g., Marxist ideology)
- Artificial human nature (from Rousseau)
- Artificial intelligence (the twenty-first century's AI)

The early modern era's need to find replacements for the foundations of the medieval world through the creation of the artificial may usefully be contemplated in the context of Scholasticism's debates about "realism" and "nominalism." To these medieval scholars, the abstract ideas above and beyond this world were *real*. No, said their opponents: those abstractions were just names; hence, nominalism. To face the language challenge, we as moderns have to consider that the Enlightenment's inventions of "artificials" were intended to shape the *real* world of *this* world. Yet they were not to be considered "real" in the Scholastic sense.

The modern era, in the early twentieth century, was elaborated upon by "modernism" in the arts, a movement into fragmentation (as in Duchamp's "Nude Descending a Staircase") that accelerated the general pressures toward fragmentation that even international law could not escape. The "international"—as an artificial concept or procedural structure that successfully could incorporate all the world's peoples in a way that bridged commonality to diversity—was fractured by the language of "modernism":

First, the *critique of representation*. In art, for example, no more landscape painting would be acceptable; the more abstract the better. In governance, this would increase pressures for direct democracy of the sort that destroyed ancient Athens.

Second, a *turn toward "primitive" sources* of cultural energy as in tribal or folk art. In governance this propelled the rise of ethnicity as providing political legitimacy.

Third, *experimentation with standards* once considered required for coherence and stability. For music, the "prepared" piano. In governance, the disregard for borders and the principles of sovereignty.

Fourth, the *juxtaposition of elements* once considered irreconcilable, such as collage or multimedia performances. In governance, providing legislative powers to administrative bureaucracies; the rise of the public-sector union, an oxymoron.[22]

Thus language in the modern age has been in a continuing process of redefinition, fragmentation, and rearrangement that has been moving toward breakdown and disputation rather than adherence to clarity and credible communication. And this phenomenon, appropriate to modernity's parallel globalization, has brought a world-spanning transformation in the commonalities of language use, largely through the ubiquity of English.

Our whole societal intercourse is fundamentally driven by communications, both society generally, and in the military chain of command. . . . Our whole set of communications now, because of these technological changes, is very much open to disruption and deception. —William J. Perry

---

The e-Revolution and the Arab World

In this context has come what appears to be the next great Revolution in human history, the e-revolution in language through which, by unprecedented technological means, any person anywhere in the world can instantly and constantly communicate with every other person.

The familiar periodization across time—Agricultural Revolution, Scientific Revolution, Industrial Revolution—was restated in its modern form by Lewis Mumford in his 1934 *Technics and Civilization*, a Big Book that attempted to make human history appear to be technology-centered. Mumford located modern technology's origins in the late Middle Ages, which he called the Eotechnic period, exemplified by the clock. Dante evokes this at the end of his *Paradiso* with the first appearance of a clock in literature, put there to say that the world would never be the same. Mumford termed the next phase Paleotechnic, which from about 1700 to 1900 would be propelled by steam. The third was Neo-technic, running on electricity. Following this taxonomy, Mumford, who died in 1990, would probably be in the vanguard of those hailing the twenty-first century as e-Technic.

While the modernist language and technology revolution is having an impact everywhere, it will produce different problems and different results in different cultures. A distinctively affected area may be the Middle East's Arab-Islamic world.

Tension between the spoken and the written word is one of the great themes of history. At least one major theory describes a fluctuation, or alternation, between two contending forms: a time of Orators (the spoken word for persuasion) and a time of Philosophers (the written word for reasoning) and then to Orators again, pendulum-like.[23] Islam is uniquely clear about this dichotomy. The Koran was spoken—dictated—by the Angel Jibril as the word of Allah to his messenger, the Prophet Mohammed, who memorized it to repeat it to his companions as scribes wrote it down. Thus the assertion that not one word of its 114 chapters, the Suras, has ever been changed.[24]

The authority of the Koran in its origin and in the incomparability of its eloquence and diction meant that spread of the faith fundamentally would be linguistic. Non-Arabic speakers must begin by learning Koranic formulas by repeating them in Arabic without understanding.[25] Thus, strictly speaking, the Koran cannot be translated. A corollary is that style and sound prevail over substance.

In the Arabic the verses are divided according to the rhythm of the language. When a certain sound that marks the rhythm recurs, there is a strong pause and the verse ends naturally, although the sentence may go on to the next verse or to several subsequent verses.[26]

Without reference to any of these characteristics of the Koran, the Arab Human Development Report, drafted by Arab intellectuals and submitted to the United Nations in 2002 under the title "Building a Knowledge Society," concluded through this perspective that Islam's spoken-word dominance affected its relationship to products of written language. Among the major "development deficits" described in the report were shortfalls in the acquisition of, and the freedom to exchange, knowledge. Put in other terms, the report was understood as saying that the Muslim fixation on the spoken Koran inclined the culture to be indifferent to the written word in non-Arabic languages. A commonly cited statistic was that Spain translated more books from non-Spanish languages in one year than the Arabic-speaking world had ever translated into Arabic in all history. The oral was authentic; the written was assumed to be tendentious or insignificant.[27]

The volcanic eruption of e-communication in the last few years has brought an exceptional array of challenges to governance in the Arab-Islamic realm. The emergence of a certain type of modern state, at once ideological and dictatorial, together with the vast extension of the technology of communication, has given propaganda wider scope and intensity. Autocratic regimes see these communicative breakthroughs as new ways to increase their powers over their peoples.

At the same time, the deeply rooted instrument of communication, the *Khutbah*, the Friday mosque sermon, once a virtual monopoly of the regime in communicating from ruler to people, and the accepted way for the latter to submit to the sovereignty of the former, has dispersed and, in extreme cases, been used by imams to rail against state authority on the grounds that simply being a "state" is to be un-Islamic.[28] The sermons of Abu Bakr al-Baghdadi, the "caliph" of the Islamic State, have been e-transmitted across the world to mobilize support and arouse passions against established order.[29] Other languages, too, have made their presence felt: terse and inflammatory messages of social media such as Twitter are "e-words" that arrive in the form of, and with the impact of, oratorical demagoguery.

Underlying this revolutionary communicative upheaval remains the fact of the Arab world's need for economic development in order to "catch up" to others, as called for in the Arab Human Development Report. After managing for centuries "to evade the revolutionary inroads of print culture," there has come the unavoidable recognition that knowledge crucial to development must be communicated through documents written in Arabic.[30]

Thus, the Arab-Islamic world, which for centuries has shaped, with excruciating care, a culture based on a clear and divinely legitimated hierarchy of the spoken over the written word, finds itself in turbulence as speeches and writings not only get in each other's way but are weapons in layers of psychological, political, and actual warfare, triggering struggles for power that societies both utilize and fear. This war of words may yet be in its early stages. It is unlikely that the Arab-Islamic culture and polity can avoid some form of world-historical transformation as a result.

It seems to me trust is an indispensable attribute of a democracy. And to the extent that trust is eroded in both institutions and in our fellow citizens, we're in trouble.   —David M. Kennedy

---

Anarchy from Below, Manipulation from Above: Disruption in Democracy

This chapter has been a review of the Enlightenment's initiation and later elaboration of a new language and a depiction of how that language has transmogrified into segments and fragments, each one of which has altered a dimension of modern discourse. The humanities and the technologies, inextricably linked by language, are locked in a struggle for control of that language. At present, technology has taken command of language to both serve technology and distort linguistic standards; one need only look at the average self-published book, blog post, or tweet for evidence. This process disrupts and corrodes the foundations of the modern era and shows no sign of being able to positively reconstruct from what it is tearing down.[31]

Modern political philosophy, with recognition of the wisdom in classical texts, attempted to devise ways in governing structures and systems to curb politically deleterious tendencies in human nature. In three significant cases, the modern approach is now being undermined by the disruptive powers of twenty-first-century language technology.

Empowering Sociopaths

Freud's *Civilization and Its Discontents* revealed that the benefits of civilized order and progress require the relegation of powerfully disruptive behaviors and desires to "the unconscious" mind. While many assume that societies primarily shape individuals, Louis Menand writes, "Freud thought that it was the other way around, that society is just a macro form of the individual, and takes its imprint from individual psychology."[32] Most today would agree that human and societal development is a two-way street, dependent on one another.

Today's social media distorts this relationship. Instant communication by way of systems such as Twitter makes it possible for individuals to immediately express the slightest emotionally disruptive and damaging reaction to events or ideas to a world-spanning audience. Opinions and private outbursts once perceived as self-harmful blunders, resolved by improving one's repressive subconscious mechanisms, are now instantly exposed to multitudes in permanent form. Civilization depends upon the time and ability to contain such eruptions; the "discontents" created by acts of self-control are the price of civil society. Were every discontent expressed, the public sphere would collapse as "all communication, all the time," instantly, produces a surrounding effect. As the astute columnist Peggy Noonan wrote, we are agitating and exciting "the unstable" sector of the population, a sector that increasingly grows larger, a Pandora's box of once subconscious partisan venom breaking open as no one becomes able to suppress the slightest discontent.

Enlarging Dictatorial Powers

As the individual is "liberated" by the ability to promulgate unconstrained feelings in every direction, the governing regimes of the world are gaining new powers of surveillance, intrusion, and control over their populations. The 2011 Arab Spring uprisings were considered at first to be made possible by the new language-spreading technologies in every young person's hand. It was widely agreed, at the time, that such tools of expression would be beyond the abilities of dictators to control. Such an assumption was foolhardy; the Arab Spring was crushed in a few short months as the old powers—colonels, hereditary monarchs, strong-armed clans with puppet "parliaments"—regained control even as they were assaulted by even more ideologically autocratic radicals claiming religious dominion.

The major one-party authoritarian regimes, too, notably Russia and the People's Republic of China, are perfecting their own domination of the new languages of disruption: techniques of interception, cooperation, blockage, elimination, falsification, and more. This reality sharply reverses earlier assumptions that major multinational corporations would be replacing states as the most potent international entities. Recent steps by the People's Republic of China to assert "cybersovereignty" bear this out. When Apple had no choice but to accept the PRC's ban on "apps" that could bypass the regime's Great Firewall of China, the power of the autocratic state over the private corporate entity was made clear to all.

This trend has begun to give authoritarian regimes unprecedented powers to suppress freedom of speech and to indoctrinate entire populations in twenty-first-century versions of Orwellian "Newspeak," such as China's propaganda that communism and capitalism are one and the same.

Disdaining the American Design: From Moderate Republic to Direct Democracy

Another recent phenomenon is the deterioration of respect traditionally given to "the deliberative process." This process, once deemed essential to the civil discourse of a polity, values balance and consensus over strident factionalism. Individuals and associations engaged in the political process were allowed the space, time, and confidentiality to examine and debate a range of options, unexposed to outside criticism, before reaching their decision and putting it before the public and the opposing party. The new language technologies, combined with crowbar-like legal methods, have made the deliberative process nearly extinct. With every individual, insider or outsider, now in effect in possession of a recording, filming, broadcasting, and publicizing piece of handheld equipment, any and all varieties of deliberative expression are so vulnerable to premature exposure that periods for careful deliberation prior to acts of decision have become rare. Equally troubling, even when such occasions are held, open discourse on policy is increasingly subject to political or legal risk.

Democracy itself, in the unique form designed by the Founders and described in Tocqueville's *Democracy in America*, is being disrupted by the new techniques of instantaneous language. To the ancient world, democracy was a tempting ideal, but understood to be dangerous, a producer of chaos that called forth a tyrant to restore order. Thucydides's Athens provided the classic case in point: swift, direct (thumbs up or down), with no patience for deliberation, and unable to prevent the deterioration of its language until "words lost their meaning." The result, as Alexander Hamilton wrote in *Federalist Papers* no. 6, was "that famous and fatal war, distinguished in the Grecian annals by the name of the Peloponnesian war; which, after various vicissitudes, intermissions, and renewals, terminated in the ruin of the Athenian commonwealth."

The result was the Founders' design for a republic that would be utterly unique: buffered against the dangers of mass decisions swiftly taken; checked and balanced, with separated powers and layered sovereignty; all within a concept of genius, *Federalist* no. 10, that would enable democracy to function effectively on a continental scale, the world's first, and still only, such example. The United States was, and still is, as Professor Samuel Huntington recognized, a "pre-modern" polity in a modern world.[33] If the modern era is ending, the United States should be better suited to manage such change than any other nation.

But not if the safeguards that make America an exceptional democracy are forfeited, lost without awareness of how or why. Yet the e-revolution can do this. The array of techniques that turn language into instantaneous power of opinion, all in the touch of a screen or a handful of words, threatens to override all the protections perfected when the republic was born.

The electronic revolution is a *language* revolution. Each of the revolutions of the modern age—French, Russian, Chinese—has brought ruination.

The world is now afflicted by an Islamist revolution, begun after the collapse of the Ottoman Empire and caliphate in the years after World War I. It produced the Islamic Republic of Iran in 1979, has been carried on by al-Qaeda and the "Islamic State and Caliphate," and is violently opposed to every element of the established modern international state system. Like all modern revolutions, it promulgates a concocted language as a weapon of power.

Only the American Revolution understood that language, like any tool or technology, must be used with care. The Founders understood that decisions made *now*, by those with power *now*, thinking only about *now*, guarantee disaster.

Understanding the inextricable centrality of language to democracy begins with the way democracy in America was designed to overcome the flaws of ancient democracy. Athens in the Periclean Age was archetypically democratic: recognized as potentially the best form of governance, but also as dangerously prone to collapse. As portrayed in Thucydides's *Peloponnesian War*, Pericles spoke proudly of Athenian democracy as swift to act by the *direct* decisions of the *demos*, the people, and unencumbered by institutions that would delay such actions. But language broke down under political, military, and societal pressures; the undeliberated decision to send a naval expedition to Sicily failed because the values of patience and foresight, the proper allocation of resources, and mature deliberation lost meaning.

The Founders of the United States knew well the story of Athens in the Peloponnesian War. They were determined that America would become a free republic, not a direct democracy. It would be a government by representatives, with dispersed sovereignty, three equal branches of government, and a variety of checks and balances.

Other political thinkers would add vitally important concepts to democracy in the modern world to overcome the problems faced in antiquity.

Kant, staying rigorously within the Enlightenment's requirement to employ "reason" alone without dependence upon outside foundational authorities, such as religion, argued his way step-by-step to demonstrate that the core of political success was *transparency* because the purpose of a state was justice, and that could only be had when the people were sovereign and could demand that their government's actions be open to examination and approval. Transparency could only truly exist in a republic, and a republic's added advantage would be that a free people would be disinclined to go to war or would hasten to end a war if war could not be avoided.

Hegel, as noted earlier, added the centrality of history, specifically "the history of the *consciousness* of freedom." In other words, history had a direction, a progression, propelled by freedom.

Tocqueville supplemented this view, seeing democracy as a force of history observable across the past several hundred years. But he knew that only if "democracy in America" is conducted wisely can democracy continue its modern trajectory.

This is a very trite way to explain it, but I'm not sure our problem is that we don't have responsive government. I think it may be that we're so responsive that we can't lead.   —Sam Nunn

---

Two concerns were paramount. First, democracy's powerful pressure is for ever-greater equality. Equality is essential, but liberty must be maintained as well so that equality does not eradicate freedom in the drive to make all outcomes equal. Second, there is, Tocqueville observed, a distinctively American democratic logic chain: religion informs mores, which inform laws, which ensure liberty, and liberty protects religion.[34] America is unique, Tocqueville said (we could also say "exceptional"), in that only in America are religion and liberty compatible; elsewhere, religion tends to suppress liberty, and liberty tends to resent and resist the demands of religion. But in America, religion sees liberty as the protector of its observances, and liberty sees religion as the cradle of its birth (as when the New England Puritan congregation was easily transposed into the town meeting). The e-revolution in communication is doing damage to this Tocquevillian narrative of American exceptionalism by making every issue "presentist" as a matter of struggles for power in current politics. If "history" appears in this battle for supremacy in current events, it is ignorantly distorted in the service of scoring power points here and now.

We May Add the United Nations

As originally chartered, the United Nations regarded democracy as only one of many forms of governance over a state; UN officials could see no reason to differentiate democracy from the other forms such as communism, fascism, and socialism. But after the end of the Cold War, the demands of many member states of the United Nations were such that the world organization recognized democracy as primarily procedural, not substantive, and therefore uniquely qualified to fit within the procedural (Westphalian) international state system itself.

* * *

There are five—there may be more—foundation stones of democracy in our time, with the United States its flagship. The changed situations in this twenty-first century almost all involve the transmogrifications of language through the e-revolution.

First, the Founders' design is being corroded by the return of direct democracy as advocated by leaders and accepted by voters unaware of democracy's history across the centuries. The attractiveness of the idea that "the more democracy the better" can doom democracy itself. The e-revolution in communication could demolish representative government. The proliferation of referenda that, when they produce unwanted results, are simply repeated—in what has been called the "never-endum"—destabilizes politics.

Kant's conclusion that transparency is the pragmatic requirement for the governmental form—a republic—that can promote peace and justice is being threatened from opposite directions simultaneously. Electronic technology has created ways—"bots"—to provide an individual with security through encryption and evasion of discovery through timed self-eradication, thus evading legal requirements for record keeping and the transparency that democratic governance is designed to ensure. At the same time, there are places and phases in political decision-making when "the deliberative process," as noted above, must be allowed to take place without public scrutiny. In recent decades, this has been violated by subpoenas, leaks, and retroactive "revelations" of who proposed what option before the deliberative process had run its course and a decision reached, ready to stand or fall in open politics. A common case would be a media headline, "Leaked Memo Reveals Smith Was Warned"; that is, a memo, one of many and various options papers, in a predecision process during which almost any argument other than the chosen decision outcome could be found to score points long after the fact. When the proprieties of deliberation are done away with, transparency paradoxically is curtailed because notes are not taken, conversations are muffled, and the historical record may be lost forever.

Second, Hegel's perception of the freedom-focused consciousness of liberty has been displaced by scientific claims (in each case disputed or problematic). "The Conscious Controversy" has done much toward dismissing the formerly assumed reality of consciousness itself as the brain/mind dichotomy has scientifically evolved into "there is no 'mind'; all is 'brain' and therefore 'consciousness' is physiological, not intellectual."[35]

The prominence of social sciences, each of which propels versions of the scientific method into realms formerly considered to be inevitably uncertain (that is, above and beyond the reach of scientific, replicable fact), is further evidence that the intellectual temper of the times is derogating the centuries-old assumption that human beings possess free will. If there is no free will, there can be no consciousness.

Third, Tocqueville's main message—if democracy as a force of history is not managed knowledgeably and wisely in America, it will lose momentum elsewhere—has now come to a turning point. Tocqueville's perceived unique compatibility between religion and liberty in America is now becoming an arena for tensions between divergent cultures. And the law's place in Tocqueville's logic chain has become tenuous with the rise of "positive" law and the endless dispute between those who regard the Constitution as a foundation upon which to base decisions and those who take it as a platform for progressively compelled social changes.

And, fourth, at the United Nations, the acceptance of democracy as a procedural part of the established international state system has declined considerably since the post–Cold War years, giving way in many places to the "model" of one-party regimes pursuing ways to benefit from economic globalization while at the same time curtailing political freedom within their borders. "The China Model" is immensely attractive to autocratic strongmen on every continent as it combines the best of all dictatorial desires: a globalized economy to skim off financial gains; political power protected by a praetorian guard; and lifelong possession of the nation's highest office. While the United Nations can still incline toward democracy promotion, it has no ability or mandate to try to reverse this trend.

Finally, underlying all this, and indeed all intellectual history, is the way a society relates language to acts. Thucydides declared that his book was "the greatest speech" about "the greatest action," the Peloponnesian War of 431–404 BC. That book narrated the ways in which Athens failed to wisely relate language to acts. The essence of the challenge is for the society through its speakers—statesmen and authors—to avoid imposing meaning on actions but rather to locate meaning in them. In a well-governed state, acts and opinions of the people will convey meanings to an alert and responsive political leader, but if leaders preemptively interpret the will of the people, good government will soon be gone.

The crisis of our time is that technics, to adopt Mumford's term, have commandeered the speakers of words. To Emerson, this is when "Things are in the saddle and ride Mankind."

My reaction is that we have documented the fact that the world ahead of us is not going to be anything like the world behind us. Change is taking place that's profound, and it's driven not by the humanities, but by technology. But we have to react to it in a humanities way—we have to think it through in human terms.   —George P. Shultz

---

AFTERWORD

James Timbie

We often struggle to find the right path forward. We all know much more work is needed. But I would combine that realism with an optimism about how we can bring together multiple viewpoints and multiple disciplines—biologists, politicians, engineers, historians, basic scientists, journalists—to take these fundamental discoveries and use them to ensure benefit to the world.
　　　　　　　　　　　　　　　　　　　—Persis Drell

Technology disrupts, and societies respond. But it is important to recognize why we pursue that disruption in the first place: it carries with it great benefit. Indeed, the authors in this book and participants at the conference on advancing technology and its impact on governance at the Hoover Institution identified many benefits of twenty-first-century technologies:

- Health outcomes and quality of life benefit from better clinical decisions, more extensive monitoring, and more emphasis on prevention made possible by early identification of risks.
- New techniques for genetic engineering and editing provide new tools for research into all living systems.

- Commerce continues to become more efficient.
- Additive manufacturing allows products to be customized for each individual at no additional cost and encourages production of products locally when and where they are needed.
- Automation increases productivity, leading to a growing GDP and better standard of living.
- Social media connects friends, families, and like-minded individuals worldwide.
- Soon autonomous cars will provide safer transportation and autonomous trucks will improve the efficiency of transport and delivery.

Of course, it's a double-edged sword.   —William J. Perry

Each of these promises to be so valuable that once in hand, societies of the future would find them intolerable to live without. And that is why those of us who live in this time of rapid technological change must navigate threats, among them those we have described:

- Tropical diseases spread by mosquitoes are moving north in response to global warming.
- Our country faces the prospect of swarms of small, cheap, but powerful weapons in the hands of state and nonstate actors, forcing us to rethink our military posture and creating potential new vulnerabilities at home that could reduce support for intervention overseas.
- The infrastructure that supports modern society is increasingly fragile and vulnerable to disruption by new cyberthreats, which are difficult to deter or to defend against.
- Perhaps half of workers are in occupations vulnerable to disruption in the near term by advancing technology.

- The benefits of economic growth fostered by advancing technologies are unevenly distributed, with more than half of total household income now going to the top 20 percent.
- The exponentially growing amount of data collected from sensors monitoring individuals and their activities raises new privacy concerns.
- Modern network technologies have a divisive polarizing effect and have been exploited by Russia to interfere with democratic elections.

There's an optimism and there's a concern that go together. If we understand these changes better, we'll be able to get the advantages and try to deal with concerns at the same time.

—John B. Taylor

We know that these twenty-first-century technologies and the challenges they pose will naturally arrive faster than governments react. But being prepared with a strategy will help. Our authors here have furthermore outlined constructive ways forward for governments and individuals to do so:

- Many pointed to the need to improve K–12 education, where the United States has fallen behind much of the rest of the world. Twenty-first-century skills depend on a good basic education. Fortunately, new technologies can be part of the solution.
- The spread of tropical diseases into more temperate areas is quite predictable, and we can prepare our diagnostic and treatment tools accordingly.
- Many noted the importance of increasing resilience—especially the resilience of our networks to cyberattack and our space assets to attacks of all kinds. Deterrence of cyberattack is difficult, in

part because of the difficulty of reliably and quickly determining attribution; hence the importance of increasing resilience rather than relying entirely on deterrence.

- With advancing technology comes disruption of many work-places. Transitions to new occupations require new skills; community college training programs have proved successful in providing new skills for new jobs. The need is to bring people and jobs together. It was noted that there are currently more than six million unfilled jobs in the United States, another demonstration of the need to bring people and jobs together.
- Advancing technology increases productivity. Given the demographics of slower growth in the working-age population, productivity gains will be essential for maintaining growth in our standard of living. Technology gains should therefore be welcome.

This conference was dedicated to the memory of Sid Drell, who participated actively and constructively in conferences just such as this, on a wide variety of subjects. If he were with us, Sid would help analyze the problems and develop practical solutions, and would advise us not to be alarmed but to work to secure the benefits of technology while dealing with its problems.

We are not gods, but that we must live with; that is what we must live with because we've created it.

—Sidney Drell

NOTES

1. Technological Change and the Workplace

1. "Artificial intelligence" is a catchy phrase, first introduced by John McCarthy at a conference at Dartmouth in 1955, that can be interpreted in a variety of ways. For this chapter, "artificial intelligence" refers to machines that can sense their environment, learn by trial and error, solve problems, and take action. This is the machine learning that enables autonomous vehicles and other means of disruption of the workplace. Such machines are built with specialized hardware and software and are trained to recognize patterns in large sets of digital data.

2. Jerry Kaplan, *Artificial Intelligence: What Everyone Needs to Know* (Oxford: Oxford University Press, 2016), xii.

3. Peter Stone, Rodney Brooks, Erik Brynjolfsson, Ryan Calo, Oren Etzioni, Greg Hager, Julia Hirschberg, Shivaram Kalyanakrishnan, Ece Kamar, Sarit Kraus, Kevin Leyton-Brown, David Parkes, William Press, AnnaLee Saxenian, Julie Shah, Milind Tambe, and Astro Teller, "Artificial Intelligence and Life in 2030," One Hundred Year Study on Artificial Intelligence: Report of the 2015–16 Study Panel, Stanford University, Stanford, CA, accessed November 5, 2017, http://ai100.stanford.edu/2016-report.

4. Stone et al., "Artificial Intelligence and Life in 2030," 25.

5. Stone et al., "Artificial Intelligence and Life in 2030," 31.

6. T. X. Hammes, "Will Technological Convergence Reverse Globalization?" *Strategic Forum* 297 (July 12, 2016): 1.

7. Hammes, "Technological Convergence," 7.

8. Stone et al., "Artificial Intelligence and Life in 2030," 38; Hammes, "Technological Convergence," 11; Carl Benedikt Frey and Michael A. Osborne, "The Future of Employment: How Susceptible Are Jobs to Computerisation?" Oxford Martin Programme on Technology and Employment, Oxford Martin School, University of Oxford, September 2013, 18, accessed November 3, 2017, www.oxfordmartin.ox.ac.uk/downloads/academic/future-of-employment.pdf.

9. Frey and Osborne, "Future of Employment," 41.

10. Frey and Osborne, "Future of Employment," 47.

11. Erik Brynjolfsson and Andrew McAfee, *The Second Machine Age: Work, Progress, and Prosperity in a Time of Brilliant Technologies* (New York: W. W. Norton and Company, 2014), 189.

12. Thomas H. Davenport and Julia Kirby, *Only Humans Need Apply: Winners and Losers in the Age of Smart Machines* (New York: Harper Collins Publishers, 2016), 77.

13. Brynjolfsson and McAfee, *The Second Machine Age*, 173.

14. Brynjolfsson and McAfee, *The Second Machine Age*, 126.

15. Brynjolfsson and McAfee, *The Second Machine Age*, 129.

16. Brynjolfsson and McAfee, *The Second Machine Age*, 130.

17. Viktor Mayer-Schönberger and Kenneth Cukier, *Big Data: A Revolution That Will Transform How We Live, Work, and Think* (New York: Eamon Dolan/Mariner Books, 2014), 2.

18. "The Internet of Things: An Overview: Understanding the Issues and Challenges of a More Connected World," Internet Society (October 2015), 32, accessed November 3, 2017, https://www.internetsociety.org/sites/default/files/ISOC-IoT-Overview-20151221-en.pdf.

19. Internet Society, "The Internet of Things," 40; Federal Trade Commission, "Protecting Consumer Privacy in an Era of Rapid Change: Recommendations for Businesses and Policymakers," March 2012, accessed November 3, 2017, https://www.ftc.gov/sites/default/files/documents/reports/federal-trade-commission-report-protecting-consumer-privacy-era-rapid-change-recommendations/120326privacyreport.pdf.

20. Federal Trade Commission, "FTC Charges Deceptive Privacy Practices in Google's Rollout of Its Buzz Social Network," news release, March 30,

2011, accessed November 3, 2017, https://www.ftc.gov/news-events/press
-releases/2011/03/ftc-charges-deceptive-privacy-practices-googles-rollout
-its-buzz; Federal Trade Commission, "Facebook Settles FTC Charges That It
Deceived Consumers by Failing to Keep Privacy Promises," news release,
November 29, 2011, accessed November 3, 2017, https://www.ftc.gov/news
-events/press-releases/2011/11/facebook-settles-ftc-charges-it-deceived-con
sumers-failing-keep.

21. European Union, General Data Protection Regulation portal, accessed
November 3, 2017, www.eugdpr.org/eugdpr.org.html.

22. Cathy O'Neil, *Weapons of Math Destruction: How Big Data Increases
Inequality and Threatens Democracy* (New York: Crown, 2016), 24, 81, 105,
144, 164.

23. Harry J. Holzer, "Labor Market Pump Is Primed—We Must Take
Advantage," Brookings Institution, August 29, 2017, accessed November 3,
2017, https://www.brookings.edu/opinions/labor-market-pump-is-primed-we
-must-take-advantage.

24. For example, career technical training programs generally do not
send their students to the English and math departments, but teach their
own classes that focus on the vocabulary, writing skills, and numerical
methods appropriate to a particular field. In some cases, English and math
are built into the technical courses, so students learn all three subjects
together.

25. California draws public attention to community college career technical education programs that achieve the metric that 90 percent or more of
students who complete the program report that their current job is close to
their field of study.

26. Mark Muro, "Adjusting to Economic Shocks Tougher than Thought,"
Brookings Institution, April 18, 2016, accessed November 3, 2017, https://
www.brookings.edu/blog/the-avenue/2016/04/18/adjusting-to-economic
-shocks-tougher-than-thought.

27. Google privacy policy, accessed April 17, 2017, https://www.google
.com/policies/privacy.

28. Network Advertising Initiative, accessed November 3, 2017, https://
www.networkadvertising.org.

29. FTC, "Protecting Consumer Privacy," ii.

30. "K-12 School Service Provider Pledge to Safeguard Student Privacy," Future of Privacy Forum and the Software and Information Industry Association, accessed November 3, 2017, https://studentprivacypledge.org/privacy -pledge.

31. Federal Trade Commission, "Data Brokers: A Call for Transparency and Accountability," May 2014, accessed November 3, 2017, https://www.ftc .gov/system/files/documents/reports/data-brokers-call-transparency -accountability-report-federal-trade-commission-may-2014/140527databroker report.pdf; "Big Data: Seizing Opportunities, Preserving Values," Executive Office of the President, May 2014, 62, accessed November 3, 2017, https:// obamawhitehouse.archives.gov/sites/default/files/docs/big_data_privacy _report_may_1_2014.pdf.

32. Ariel Ezrachi and Maurice E. Stucke, *Virtual Competition: The Promise and Perils of the Algorithm-Driven Economy* (Cambridge, MA: Harvard University Press, 2016), 95.

33. FTC, "Protecting Consumer Privacy," viii.

34. Gary S. Miliefsky, "The Top Ten Mobile Flashlight Applications Are Spying on You. Did You Know?" *Cyber Defense Magazine*, October 1, 2014, accessed November 3, 2017, www.cyberdefensemagazine.com/the-top-ten -mobile-flashlight-applications-are-spying-on-you-did-you-know.

2. Technological Change and the Fourth Industrial Revolution

1. Klaus Schwab, *The Fourth Industrial Revolution: What It Means, How to Respond* (Geneva, Switzerland: World Economic Forum, 2016), accessed February 15, 2016, https://www.weforum.org/agenda/2016/01/the-fourth -industrial-revolution-what-it-means-and-how-to-respond.

2. "Turning Their Backs on the World," *The Economist*, February 19, 2009, accessed April 26, 2016, www.economist.com/node/13145370.

3. *The Economist*, "Turning Their Backs on the World."

4. World Bank, "Merchandise Trade (% of GDP)," accessed December 28, 2016, http://data.worldbank.org/indicator/TG.VAL.TOTL.GD.ZS/countries?dis play=graph; World Bank, "Global Growth Edges Up to 2.7 Percent Despite Weak Investment," news release, January 10, 2017, accessed January 13, 2017, www.worldbank.org/en/news/press-release/2017/01/10/global-growth-edges -up-to-2–7-percent-despite-weak-investment.

5. World Bank, "Trade (% of GDP)," accessed December 28, 2016, http://data.worldbank.org/indicator/NE.TRD.GNFS.ZS/countries/1W-CN-US?display=graph.

6. World Bank, "Merchandise Exports (Current US$)," accessed March 20, 2017, http://data.worldbank.org/topic/trade.

7. Matthieu Bussiere, Julia Schmidt, and Natacha Valla, *International Financial Flows in the New Normal: Key Patterns (and Why We Should Care)*, CEPII (Centre d'Études Prospectives et d'Informations Internationales), March 2016, accessed May 26, 2016, www.cepii.fr/PDF_PUB/pb/2016/pb2016-10.pdf.

8. Ruchir Sharma, *The Rise and Fall of Nations: Forces of Change in the Post-Crisis World* (New York: W. W. Norton, 2016), 5.

9. "United States: Foreign Investment," Santander Trade Portal, accessed July 10, 2017, https://en.portal.santandertrade.com/establish-overseas/united-states/foreign-investment.

10. US Commerce Department, Economics & Statistics Administration, "Foreign Direct Investment in the United States: Update to 2013 Report," accessed July 10, 2017, http://esa.gov/reports/foreign-direct-investment-united-states-update-2013-report.

11. Rana Foroohar, "We've Reached the End of Global Trade," *Time*, October 12, 2016, accessed January 15, 2017, http://time.com/4521528/2016-election-global-trade.

12. Kristin Forbes, "Financial 'Deglobalization'? Capital Flows, Banks, and the Beatles," speech delivered at Queen Mary University, London, November 28, 2014.

13. Simon Nixon, "Risk of Deglobalization Hangs over World Economy," *Wall Street Journal*, October 5, 2016, accessed December 4, 2016, www.wsj.com/articles/risk-of-deglobalization-hangs-over-world-economy-14756 85469.

14. Susan Lund, James Manyika, and Jacques Bughin, "Globalization Is Becoming More about Data and Less about Stuff," *Harvard Business Review*, March 14, 2016, accessed December 28, 2016, https://hbr.org/2016/03/globalization-is-becoming-more-about-data-and-less-about-stuff.

15. James Manyika, Susan Lund, Jacques Bughin, Jonathan Woetzel, Kalin Stamenov, and Dhruv Dhingra, "Digital Globalization: The New Era of Global Flows," McKinsey Global Institute, February 2016, accessed January 12, 2017,

www.mckinsey.com/business-functions/digital-mckinsey/our-insights/digi
tal-globalization-the-new-era-of-global-flows.

16. World Bank, "GDP Per Capita (Current US$)," accessed February 5, 2017, http://data.worldbank.org/indicator/NY.GDP.PCAP.CD.

17. "The Zettabyte Era: Trends and Analysis," Cisco, June 2, 2016, accessed February 16, 2017, www.cisco.com/c/en/us/solutions/collateral/service-pro vider/visual-networking-index-vni/vni-hyperconnectivity-wp.html.

18. Paul Davidson, "More Robots Coming to U.S. Factories," *USA Today*, February 10, 2015, accessed April 18, 2016, www.usatoday.com/story/money /2015/02/09/bcg-report-on-factory-robots/23143259.

19. "Wages and Employment," *China Labour Bulletin*, accessed April 26, 2015, www.clb.org.hk/content/wages-and-employment.

20. April Glaser, "The U.S. Will Be Hit Worse by Job Automation than Other Major Economies," *Recode*, March 25, 2017, accessed May 22, 2017, https://www.recode.net/2017/3/25/15051308/us-uk-germany-japan-robot-job -automation.

21. Brian Wang, "Annual World Industrial Robot Installations Expected to Double from 2014 to 2019," *NextBigFuture*, April 4, 2016, accessed April 5, 2016, http://nextbigfuture.com/2016/04/annual-world-industrial-robot.html.

22. Peggy Hollinger, "Meet the Cobots: Humans and Robots Work Together on the Factory Floor," *Financial Times*, May 5, 2016, accessed May 5, 2016, https://next.ft.com/content/6d5d609e-02e2-11e6-af1d-c47326021344.

23. Vinod Baya and Lamont Wood, "Service Robots: The Next Big Productivity Platform," *PwC Technology Forecast* 2 (2015): 9, accessed April 5, 2016, www.pwc.com/us/en/technology-forecast/2015/robotics/features/assets/ 26769-2015-tech-forecast-robotics-article-1-v6.pdf.

24. Nicholas St. Fleur, "3-D Printing Just Got 100 Times Faster," *The Atlantic*, March 17, 2015, accessed July 17, 2017, https://www.theatlantic.com/ technology/archive/2015/03/3d-printing-just-got-100-times-faster/388051.

25. "ORNL, Cincinnati Partner to Develop Commercial Large-Scale Additive Manufacturing System," Oak Ridge National Laboratory, February 17, 2014, accessed April 3, 2016, https://www.ornl.gov/news/ornl-cincinnati-part ner-develop-commercial-large-scale-additive-manufacturing-system.

26. Laura Cox, "Rapid Liquid Printing: 3D Printing 2.0," *D/sruption*, May 17, 2017, accessed May 22, 2017, https://disruptionhub.com/3d-printing -evolves-rapid-liquid-printing.

27. Eddie Krassenstein, "CloudDDM—Factory with 100 (Eventually 1,000) 3D Printers & Just 3 Employees Opens at UPS's Worldwide Hub," *3DPrint.com*, May 4, 2015, accessed April 3, 2016, https://3dprint.com/62642/cloudddm-ups.

28. "UPS to Launch On-Demand 3D Printing Manufacturing Network," *UPS Pressroom*, May 18, 2016, accessed July 25, 2017, https://www.pressroom.ups.com/pressroom/ContentDetailsViewer.page?ConceptType=Press Releases&id=1463510444185-310.

29. "3D Printing Comes of Age in US Industrial Manufacturing," Price Waterhouse Cooper in conjunction with Manufacturing Institute, April 2016: 1, accessed April 21, 2016, www.pwc.com/us/en/industrial-products/publications/assets/pwc-next-manufacturing-3d-printing-comes-of-age.pdf.

30. Josh Zumbrun, "Forces That Opened Up Borders Show Signs of Sputtering," *Wall Street Journal*, April 4, 2016.

31. Nanette Byrnes, "AI Hits the Mainstream," *MIT Technology Review* 119, no. 3 (May/June 2016): 62.

32. Melissa Korn, "There's a Reason the Teaching Assistant Seems Robotic," *Wall Street Journal*, May 7, 2016.

33. Susan Beck, "AI Pioneer ROSS Intelligence Lands Its First Big Law Clients," *The American Lawyer,* May 6, 2016, accessed May 12, 2016, www.americanlawyer.com/id=1202757054564/AI-Pioneer-ROSS-Intelligence-Lands-Its-First-Big-Law-Clients?slreturn=20160412132159.

34. In figure 2.5, occupations are categorized as follows. Nonroutine cognitive includes BLS categories "management, professional, and related" occupations; routine cognitive includes "sales and office" occupations; nonroutine manual includes "service" occupations; while routine manual includes "production, transportation and material moving," "installation, maintenance, and repair," and "construction and extraction" occupations.

35. Scott Santens, "Deep Learning Is Going to Teach Us All the Lesson of Our Lives: Jobs Are for Machines," *Basic Income*, March 16, 2016, accessed May 25, 2016, https://medium.com/basic-income/deep-learning-is-going-to-teach-us-all-the-lesson-of-our-lives-jobs-are-for-machines-7c6442 e37a49?imm_mid=0e3eb5&cmp=em-na-na-na-newsltr_econ_20160520#.f7fjco7r3.

36. "Renewables 2015: Global Status Report," Ren 21: Renewable Energy Policy Network for the 21st Century, July 2015: 17, accessed May 5, 2016,

www.ren21.net/wp-content/uploads/2015/07/REN12-GSR2015_Onlinebook
_low1.pdf.

37. Joe Ryan, "A Renewables Revolution Is Toppling the Dominance of Fossil Fuels in U.S. Power," *Bloomberg*, February 4, 2016, accessed May 5, 2016, www.bloomberg.com/news/articles/2016-02-04/renewables-top-fossil-fuels -as-biggest-source-of-new-u-s-power.

38. Fred Lambert, "Electric Vehicle Sales to Surpass Gas-Powered Cars by 2040, Says New Report," *Electrek,* May 5, 2017, accessed July 7, 2017, https:// electrek.co/2017/05/05/electric-vehicle-sales-vs-gas-2040.

39. U.S. Energy Information Administration, "U.S. Energy Imports and Exports to Come into Balance for First Time since 1950s," April 15, 2015, accessed May 5, 2016, www.eia.gov/todayinenergy/detail.cfm?id=20812.

40. Scott Malcomson, *Splinternet: How Geopolitics and Commerce Are Fragmenting the World Wide Web* (New York: OR Books, 2016).

41. Simon Denyer, "China's Scary Lesson to the World: Censoring the Internet Works," *Washington Post*, May 23, 2016, accessed November 3, 2017, https://www.washingtonpost.com/world/asia_pacific/chinas-scary-lesson -to-the-world-censoring-the-internet-works/2016/05/23/413afe78-fff3-11e5 -8bb1-f124a43f84dc_story.html?utm_term=.8908acc0a7de.

42. Kevin Lui, "China Just Made It Even Harder to Get Around the Great Firewall," *Time*, January 23, 2017, accessed May 22, 2017, http://time.com/ 4642916/china-vpn-internet-great-firewall-censorship.

43. "Global Public Opinion in the Bush Years (2001–2008)," Pew Research Center, December 18, 2008, accessed April 29, 2016, www.pewglobal.org/2008 /12/18/global-public-opinion-in-the-bush-years-2001-2008/#enthusiasm-for -globalization.

44. Bruce Stokes, "Most of the World Supports Globalization in Theory, but Many Question It in Practice," Pew Research Center, September 16, 2014, accessed April 29, 2016, www.pewresearch.org/fact-tank/2014/09/16/most-of -the-world-supports-globalization-in-theory-but-many-question-it-in-practice.

45. Fred Pleitgen and James Masters, "Merkel Gets a Fourth Term but German Voters Deliver Far-Right Surge," *CNN*, September 25, 2017, accessed November 3, 2017, www.cnn.com/2017/09/24/europe/german-election-results/ index.html.

46. Zumbrun, "Forces That Opened Up Borders."

47. Paul Hannon, "OECD Indicators Signal Stronger Economic Growth," *Wall Street Journal*, January 11, 2017.

48. Roberto Azevedo, "Trade Policy Review Body: Annual Overview of Developments in the International Trading Environment," World Trade Organization, December 9, 2016, accessed January 26, 2017, https://www.wto.org/english/News_e/spra_e/spra152_e.htm.

49. Nixon, "Risk of Deglobalization."

50. "The Year of Living Dangerously," *The Economist*, December 24, 2016: 11.

51. Andrzej W. Miziolek, "Nanoenergetics: An Emerging Technology Area of National Importance," *AMPTIAC Quarterly* 6, no. 1 (Spring 2002): 45, accessed November 3, 2017, http://ammtiac.alionscience.com/pdf/AMPQ6_1ART06.pdf.

52. Louis A. Del Monte, *Nanoweapons: A Growing Threat to Humanity* (Lincoln, NE: Potomac Books, 2017), 47–48.

53. James K. Sanborn, "Beacon Improves UAV's Cargo-Delivery Accuracy," *Marine Corps Times*, July 8, 2012.

54. Vision-Based Control and Navigation of Small, Lightweight UAVs, workshop, Congress Center, Hamburg, Germany, 2015, accessed September 4, 2015, www.seas.upenn.edu/~loiannog/workshopIROS2015uav; Alan Phillis, "Drones at CES 2015 Showcase UAV Technology's Bright Future," Dronelife.com, January 14, 2015, accessed September 4, 2015, http://dronelife.com/2015/01/14/drones-ces-2015-showcase-uav-technologys-bright-future.

55. Evan Ackerman and Celia Gorman, "This Drone Uses a Smartphone for Eyes and a Brain," *IEEE Spectrum*, April 9, 2015, accessed November 5, 2015, http://spectrum.ieee.org/video/robotics/aerial-robots/this-drone-uses-a-smartphone-for-a-brain.

56. Agriculture UAV Drones, Photos—Multi-Spectral Camera," Homeland Surveillance & Electronics LLC, accessed August 6, 2014, www.agricultureuavs.com/photos_multispectral_camera.htm.

57. "Future Weapons: Explosively Formed Penetrator (EFP)," *Future Weapons TV*, June 24, 2011.

58. Bill Roggio, "Troops Find EFP Factory in Sadr City," *Long War Journal*, October 30, 2008, accessed August 15, 2015, www.longwarjournal.org/archives/2008/10/iraqi_troops_find_ef.php.

59. Eddie Krassenstein, "Plus-MFg's +1000k Multi Material Metal 3D Printer Shows Its Power," *3dprint*, accessed November 3, 2017, http://3dprint .com/87236/plus-mfg-3d-metal-printer.

60. Flexrotor capabilities, accessed July 26, 2017, http://aerovelco.com /flexrotor.

61. "Official Launch of Defiant Labs and the DX-3," *The Sky Guys*, accessed November 5, 2017, http://theskyguys.ca/2016/12/official-launch-of-defiant -labs-and-the-dx-3.

62. Jordan Golson, "A Military-Grade Drone That Can Be Printed Any-where," *Wired*, September 16 2014, accessed September 4, 2015, www.wired .com/2014/09/military-grade-drone-can-printed-anywhere.

63. UAV/UCAV, *Chinese Military Aviation*, accessed April 24, 2014, http:// chinese-military-aviation.blogspot.com/p/uav.html.

64. "Switchblade," AeroEnvironment; "New IDF: Tactical Kamikaze Drones," *Arutz Sheva,* May 28, 2015, accessed November 3, 2015, www.israel nationalnews.com/News/News.aspx/195973#.VjpSSyuTSW4.

65. Robert Hewson, "Concealed-Carriage Club-K Changes Cruise Missile Rules," *Jane's Defence Weekly* 47, no. 15 (April 14, 2010): 5.

66. Mark Thompson, "The Navy's Amazing Ocean-Powered Underwater Drone," *Time*, December 22, 2013, accessed September 4, 2015, http://swamp land.time.com/2013/12/22/navy-underwater-drone.

67. "Autonomous Submarine Drones: Cheap, Endless Patrolling," Center for International Maritime Security, June 5, 2014, accessed September 5, 2015, http://cimsec.org/autonomous-subarine-drones-cheap-endless-patrolling; Will Connors, "Underwater Drones Are Multiplying Fast," *Wall Street Journal*, June 24, 2013, accessed September 4, 2015, www.wsj.com/articles/SB1000142 4127887324183204578565460623922952.

68. Andrew S. Erickson et al., *Chinese Mine Warfare: A PLA Navy's 'Assassin's Mace' Capability*, US Naval War College, China Maritime Studies no. 3: 14.

69. Thomas Shugart and Javier Gonzalez, "First Strike: China's Missile Threat to U.S. Bases in Asia," Center for a New American Security, June 28, 2017: 13.

70. James Drew, "F-35A Cost and Readiness Data Improves in 2015 as Fleet Grows," *FlightGlobal*, February 2, 2016, accessed February 15, 2017,

https://www.flightglobal.com/news/articles/f-35a-cost-and-readiness-data
-improves-in-2015-as-fl-421499.

71. Naval Air Systems Command: Tomahawk, accessed May 30, 2014,
www.navair.navy.mil/index.cfm?fuseaction=home.display&key
=F4E98B0F-33F5-413B-9FAE-8B8F7C5F0766.

72. Tyler Rogoway, "More Details Emerge on Kratos' Optionally
Expendable Air Combat Drones," *The Warzone*, February 7, 2017, accessed
February 14, 2017, www.thedrive.com/the-war-zone/7449/more-details-on
-kratos-optionally-expendable-air-combat-drones-emerge.

73. Aaron Gregg, "Robotic Wingmen May Fly Next to Fighter Jets," *Washington Post*, June 15, 2017.

74. Mark Selinger, "NGA Growing in Acceptance of Satellite Imagery
Startups," *viaSatellite*, September 28, 2016, accessed August 3, 2017, www
.satellitetoday.com/newspace/2016/09/28/nga-growing-acceptance-satellite
-imagery-startups.

75. Robert Work, "The Third U.S. Offset Strategy and Its Implications for
Partners and Allies," January 28, 2015, accessed November 22, 2015, www
.defense.gov/News/Speeches/Speech-View/Article/606641/the-third-us
-offset-strategy-and-its-implications-for-partners-and-allies.

76. Eliot A. Cohen, "Global Challenges, U.S. National Security Strategy,
and Defense Organization," testimony to Senate Armed Services Committee,
October 22, 2015, accessed November 6, 2015, www.armed-services.senate
.gov/imo/media/doc/Cohen_10-22-15.pdf.

77. GDP Ranking for 2016, World Bank, accessed July 18, 2017, https://
data.worldbank.org/data-catalog/GDP-ranking-table.

78. "The World Factbook," Central Intelligence Agency, accessed July 18,
2017, https://www.cia.gov/library/publications/the-world-factbook/rankorder/
2004rank.html.

79. United Nations Conference on Trade and Development, "World
Investment Report 2017," 225, accessed July 18, 2017, http://unctad.org/en/
PublicationsLibrary/wir2017_en.pdf.

80. Joel Kotkin, "Death Spiral Demographics: The Countries Shrinking the
Fastest," *Forbes*, February 1, 2017, accessed July 14, 2017, https://www.forbes
.com/sites/joelkotkin/2017/02/01/death-spiral-demographics-the-countries
-shrinking-the-fastest/#5fbad012b83c.

81. "Russia Caught between Economic Decline and Potentially Explosive Demographic Change," *Johnson's Russia List*, July 6, 2017, accessed July 28, 2017, http://russialist.org/russia-caught-between-economic-decline -and-potentially-explosive-demographic-change.

82. Ivana Kottasova, "Russia Is Planning for Low Oil Prices for Years," *CNN Money*, October 14, 2016, accessed November 5, 2017, http://money.cnn .com/2016/10/14/news/economy/russia-budget-oil-price/index.html.

83. Mark Galeotti, "The Truth about Russia's Defence Budget," European Council on Foreign Relations, March 24, 2017, accessed July 13, 2017, www .ecfr.eu/article/commentary_the_truth_about_russias_defence_budget _7255.

84. Robbie Whelan and Esther Fung, "China's Factories Count on Robots as Workforce Shrinks," *Wall Street Journal*, August 17, 2016.

85. Conner Forrest, "Chinese Factory Replaces 90% of Humans with Robots, Production Soars," *TechRepublic*, July 30, 2015, accessed November 5, 2017, www.techrepublic.com/article/chinese-factory-replaces-90-of-humans -with-robots-production-soars.

86. Forrest, "Chinese Factory."

87. Mandy Zuo, "Rise of the Robots: 60,000 Workers Culled from Just One Factory as China's Struggling Electronics Hub Turns to Artificial Intelligence," *South China Morning Post*, May 21, 2016, accessed December 15, 2016, http://fortune.com/2016/05/25/adidas-robot-speedfactories.

88. Ben Blanchard and John Ruwitch, "China Hikes Defense Budget, to Spend More on Internal Security," *Reuters*, March 4, 2013, accessed August 8, 2017, www.reuters.com/article/us-china-parliament-defence-idUSBRE92403 620130305.

89. Kotkin, "Death Spiral Demographics."

90. See T. X. Hammes, "Offshore Control: A Proposed Strategy for an Unlikely Conflict," *Institute for National Strategic Studies*, June 2012, accessed December 2, 2016, www.dtic.mil/dtic/tr/fulltext/u2/a577602.pdf; and Andrew F. Krepinevich, Jr., "How to Deter China: The Case for Archipelagic Defense," *Foreign Affairs*, March/April 2015, accessed December 2, 2016, https://www .foreignaffairs.com/articles/china/2015-02-16/how-deter-china.

91. Thomas Shugart, "First Strike: China's Missile Threat to U.S. Bases in Asia," Center for a New American Security, June 28, 2017, accessed August 8, 2017, https://www.cnas.org/publications/reports/first-strike-chinas-missile -threat-to-u-s-bases-to-asia.

92. Schwab, "The Fourth Industrial Revolution."

93. Bruce Drake and Carroll Doherty, "Key Findings on How Americans View the U.S. Role in the World," Pew Research Center, May 5, 2016, accessed February 23, 2017, www.pewresearch.org/fact-tank/2016/05/05/key-findings -on-how-americans-view-the-u-s-role-in-the-world.

94. Harriet Torry, "U.S. Factor-Sector Activity Hits 13-Year High," *Wall Street Journal*, October 2, 2017.

3. Governance and Security through Stability

1. Bernard Brodie, ed., *The Absolute Weapon: Atomic Power and World Order*, February 16, 1946, draft, Eisenhower Library, 1946, accessed November 4, 2017, https://www.osti.gov/opennet/servlets/purl/16380564-wvLB09/16 380564.pdf; Michael Quinlan, *Thinking about Nuclear Weapons: Principles, Problems, Prospects* (Oxford: Oxford University Press, 2009).

2. Explosive energy release and potential deaths if used in a populated area can both be more than 10^4 to 10^6 times larger for nuclear than for conventional explosions.

3. Elbridge A. Colby and Michael S. Gerson, eds., *Strategic Stability: Contending Interpretations* (Carlisle, PA: Strategic Studies Institute and Army War College Press, 2013), accessed November 3, 2017, http://ssi.armywarcollege.edu /pdffiles/pub1144.pdf.

4. The speed of computing (e.g., floating operations per second, flops), the number density of transistors (Moore's Law), and the number of internet users have all seen exponential growth as a function of time over the past decades. Smartphones represent a qualitative shift in the ubiquity of computing as well as communications worldwide; National Research Council (NRC), *At the Nexus of Cybersecurity and Public Policy: Some Basic Concepts and Issues* (Washington, DC: National Academies Press, 2014), accessed November 3, 2017, https://www.nap.edu/18749.

5. Rita Tehan, *Cybersecurity: Critical Infrastructure Authoritative Reports and Resources*, Congressional Research Service Report R44410, 2017, accessed November 3, 2017, https://fas.org/sgp/crs/misc/R44410.pdf.

6. United Nations Group of Governmental Experts, "Developments in the Field of Information and Telecommunications in the Context of International Security," report A/70/174, United Nations General Assembly, 2015, accessed November 3, 2017, http://undocs.org/A/70/174.

7. Stuart J. Russell and Peter Norvig, *Artificial Intelligence: A Modern Approach*, 3rd ed. (Upper Saddle River, NJ: Prentice-Hall, 2010); Peter Stone, Rodney Brooks, Erik Brynjolfsson, Ryan Calo, Oren Etzioni, Greg Hager, Julia Hirschberg, Shivaram Kalyanakrishnan, Ece Kamar, Sarit Kraus, Kevin Leyton-Brown, David Parkes, William Press, AnnaLee Saxenian, Julie Shah, Milind Tambe, and Astro Teller, "Artificial Intelligence and Life in 2030," One Hundred Year Study on Artificial Intelligence: Report of the 2015–16 Study Panel, Stanford University, Stanford, CA, accessed November 5, 2017, http://ai100.stanford.edu/2016-report.

8. Although we tend toward technological optimism, and therefore assume that many of the current hurdles may be overcome through further research, we make the comparison with nonlinear chaotic systems that exhibit fundamentally unpredictable behavior.

9. Stuart Russell, Daniel Dewey, and Max Tegmark, "Research Priorities for Robust and Beneficial Artificial Intelligence," *AI Magazine* 36 (2015), 105–14, accessed November 3, 2017, https://futureoflife.org/data/documents /research_priorities.pdf.

10. National Academies of Sciences, Engineering, and Medicine, *Human Genome Editing: Science, Ethics, and Governance* (Washington, DC: National Academies Press, 2017), accessed November 4, 2017, https://www.nap .edu/24623.

11. Multiple uses include benefits for medicine, agriculture, and other domains; commercial use; and harmful applications in either a military or terrorist context.

12. Terrence M. Tumpey, C. F. Basler, P. V. Aguilar, H. Zeng, A. Solórzano, D. E. Swayne, N. J. Cox, J. M. Katz, J. K. Taubenberger, P. Palese, and A. Garcia-Sastre, "Characterization of the Reconstructed 1918 Spanish Influenza Pandemic Virus," *Science* 310 (2005): 77–80.

13. David Livingstone and Patricia Lewis, *Space, the Final Frontier for Cybersecurity?* (London: Chatham House, Royal Institute of International Affairs, 2016), accessed November 3, 2017, https://www.chathamhouse.org /publication/space-final-frontier-cybersecurity; National Academies of Sciences, Engineering, and Medicine, *National Security Space Defense and Protection: Public Report* (Washington, DC: National Academies Press, 2016), accessed November 3, 2017, www.nap.edu/23594.

14. Defense Science Board, *Summer Study on Autonomy*, Department of Defense, Washington, DC, 2016, accessed November 3, 2017, www.acq.osd .mil/dsb/reports/2010s/DSBSS15.pdf; T. X. Hammes, "Technologies Converge and Power Diffuses: The Evolution of Small, Smart, and Cheap Weapons," Policy Analysis 786, Cato Institute, 2016, accessed November 4, 2017, https:// www.cato.org/publications/policy-analysis/technologies-converge.

15. The formulation could be "to prevent nuclear war," but we consider this as implying more control over future events than can be reasonably expected.

16. Civil wars, noted for their ferocity and destructiveness, profoundly change and often destroy the societies within which they take place. The scenarios being considered here can be thought of as having consequences similar to those of extended civil war, but taking place in a much shorter period of time (e.g., weeks instead of years). In addition, these scenarios assume external attack in contrast to internal conflict, though the distinction is necessarily blurred when external forces lend support to one or the other antagonist in a civil war.

17. Richard L. Garwin, "Strategic Security Challenges for 2017 and Beyond," presentation given May 1, 2017, at the National Academy of Sciences, Washington, DC, accessed November 3, 2017, https://fas.org/rlg/nas-challenges.pdf.

18. We admit that this is a doubly speculative thesis, both because there is no experience with the actual use of modern nuclear arsenals in war (as distinct from their role for deterrence) and because the consequences of other technologies' catastrophic misuses remain largely speculative at present. For a different perspective, see Jeremy Rabkin and John Yoo, *Striking Power: How Cyber, Robots, and Space Weapons Change the Rules for War* (New York: Encounter Books, 2017).

19. We presume that crises cannot be avoided altogether, and focus on those crises that have the potential for catastrophic outcome.

20. See also chapters 3 and 4, James E. Goodby, *Approaching the Nuclear Tipping Point: Cooperative Security in an Era of Global Change* (Lanham, MD: Rowman and Littlefield, 2017).

21. Assured second strike—the ability to counter after a nuclear attack—provides an ultimate deterrent for several nuclear-weapon states and is currently provided by their submarine-launched nuclear missiles. Thus, nuclear deterrence would be profoundly affected should any new technologies, alone or in combination, make it possible to track these submarines.

22. Joseph S. Nye, Jr., "Nuclear Lessons for Cyber Security?" *Strategic Studies Quarterly* 5, no. 4 (2011): 18–38, accessed November 4, 2017, https://dash.harvard.edu/handle/1/8052146.

23. See Frank Pabian, "Commercial Satellite Imagery as an Evolving Open-Source Verification Technology: Emerging Trends and Their Impact for Nuclear Nonproliferation Analysis," European Energy Community Joint Research Center Technical Report EUR27687, 2015. Even without loss of space capability, a government can benefit from the commercial availability of satellite services if these allow it to publicly reveal and discuss potential treaty violations, or other illegal activities, without needing to make reference to its own (possibly sensitive) capabilities. Nuclear-explosion monitoring is similarly facilitated by the large number of seismic stations deployed worldwide for purposes of earthquake monitoring and academic research, such that a government need not depend on or refer to its own data in openly identifying and characterizing suspected nuclear-explosion tests.

24. Robert H. Latiff, *Future War: Preparing for the New Global Battlefield* (New York: Knopf, 2017).

25. US Department of Energy, "The Manhattan Project: An Interactive History," accessed November 4, 2017, https://www.osti.gov/opennet/manhattan-project-history/Events/1939–1942/einstein_letter.htm.

26. Jim Mattis, "A Military Perspective," in *Andrei Sakharov: The Conscience of Humanity*, Sidney D. Drell and George P. Shultz, eds. (Stanford, CA: Hoover Institution Press, 2015).

27. Nye, "Nuclear Lessons for Cyber Security?"; Joseph S. Nye, Jr., "Deterrence and Dissuasion in Cyberspace," *International Security* 41, no. 3 (Winter 2016/17): 44–71.

28. We use "globalization" in the more general, popular sense to include all manner of shared influences among nations, from cultural and political to economical.

4. Governance in Defense of the Global Operating System

1. "Internationalism is . . . described as the theory and practice of transnational or global cooperation. As a political ideal, it is based on the belief that nationalism should be transcended because the ties that bind people of

different nations are stronger than those that separate them." N. D. Arora, *Political Science* (New York: Tata McGraw-Hill Education, 2011), 2.

2. Christian Lange, "Internationalism," Nobel Peace Prize lecture, December 13, 1921, accessed November 5, 2017, www.nobelprize.org/nobel _prizes/peace/laureates/1921/lange-lecture.html.

3. Lange, "Internationalism."

4. Alfred Tennyson, *Poems* (Boston: W. D. Ticknor, 1842).

5. Tim Kaine, "A New Truman Doctrine: Grand Strategy in a Hyperconnected World," *Foreign Affairs* 96 (July/August 2017): 4, 43.

6. Henry Kissinger, *World Order* (New York: Penguin Press, 2014), 3.

7. Kissinger, *World Order*, 4.

8. Richard Haass, *A World in Disarray: American Foreign Policy and the Crisis of the Old Order* (New York: Penguin Press, 2017), 24.

9. Dale C. Copeland, *Economic Interdependence and War* (Princeton, NJ: Princeton University Press, 2015), 4.

10. Haass, *World in Disarray*, 31.

11. Haass, *World in Disarray*, 63.

12. Haass, *World in Disarray*, 72.

13. Raymond F. DuBois, *Science, Technology and U.S. National Security Strategy: Preparing Military Leadership for the Future* (Washington, DC: Center for Strategic & International Studies, 2017).

14. Timothy Walton, "Securing the Third Offset Strategy: Priorities for the Next Secretary of Defense," *Joint Forces Quarterly* 82 (July 2016).

15. Evan Thomas, "Defensive about Defense," *Time*, March 10, 1986.

16. Daniel Kahneman, *Thinking Fast and Slow* (New York: Farrar, Straus and Giroux, 2011), 231–32.

17. Shulamis Frieman, *Who's Who in the Talmud* (Lanham, MD: Jason Aronson, 2000), 163.

18. Kaine, "A New Truman Doctrine."

19. National Academies of Sciences, Engineering, and Medicine, *National Security Space Defense and Protection: Public Report* (Washington, DC: The National Academies Press, 2016), accessed November 5, 2017, https://doi .org/10.17226/23594.

20. Obama White House Archive, "Five Things to Know: The Administration's Priorities on Cyber Security," February 12, 2013.

21. Walter Russell Mead, "What Truman Can Teach Trump," *Wall Street Journal*, July 21, 2017.

22. Henry Kissinger, Stenographic Transcript before the Committee on Armed Services, United States Senate, January 29, 2015.

23. Center for a New American Security, "Technology and National Security," accessed September 5, 2017, https://www.cnas.org/research/technology -and-national-security.

24. Jim Collins, *Good to Great: Why Some Companies Make the Leap . . . and Others Don't* (New York: HarperCollins, 2001).

25. Robert Kagan, "Backing into World War III," *Foreign Policy*, February 3, 2017: 1.

26. Kagan, "Backing into World War III," 8.

27. George P. Shultz, *Issues on My Mind: Strategies for the Future* (Stanford, CA: Hoover Institution Press, 2013).

28. McGeorge Bundy, William J. Crowe, and Sidney D. Drell, *Reducing Nuclear Danger: The Road Away from the Brink* (New York: Council on Foreign Relations Press, 1993), 12.

5. Technological Change and Global Biological Disequilibrium

1. Elizabeth Kolbert, *The Sixth Extinction: An Unnatural History* (New York: Henry Holt & Company, 2014).

2. *Restoring the Foundation: The Vital Role of Research in Preserving the American Dream* (Cambridge, MA: American Academy of Arts & Sciences, 2014).

6. Governance and Order in a Networked World

1. Philip Zelikow, "Is the World Slouching toward a Grave Systemic Crisis?" *The Atlantic*, August 11, 2017.

2. Much popular literature draws a confusing distinction between hierarchies and networks. Most networks are hierarchical in some respects, if only because some nodes are more central than others, while hierarchies are just special kinds of networks in which flows of information or resources are restricted to certain edges in order to maximize the centrality of the ruling node. The correct distinction is between hierarchical networks and distributed networks. For a detailed exploration of these two concepts, which are

not dichotomous, see Niall Ferguson, *The Square and the Tower: Networks and Power, from the Freemasons to Facebook* (New York: Penguin Press, 2018).

3. The original essay on this theme was by Niall Ferguson and Moritz Schularick, "Chimerical? Think Again," *Wall Street Journal*, February 5, 2007. We revisited it in " 'Chimerica' and the Rule of Central Bankers," *Wall Street Journal*, August 27, 2015. The idea inspired Lucy Kirkwood's 2013 play, *Chimerica*.

4. So far as I am aware, this has never been done. Relevant data can be found at the Observatory of Economic Complexity, http://atlas.media.mit .edu, accessed December 7, 2017, and http://globe.cid.harvard.edu, accessed November 4, 2017.

5. James Manyika (McKinsey Global Institute), "Playing to Win: The New Global Competition for Corporate Profits," *LinkedIn*, October 20, 2015, 11.

6. See, for example, George A. Barnett, ed., *Encyclopedia of Social Networks*, vol. 1 (Los Angeles and London: SAGE Publications, 2011), 297. The optimistic case is laid out by Anne-Marie Slaughter, *The Chessboard and the Web: Strategies of Connection in a Networked World* (New Haven, CT: Yale University Press, 2017).

7. Barnett, *Encyclopedia of Social Networks*, 371.

8. Steven Pinker and Andrew Mack, "The World Is Not Falling Apart," *Slate*, December 22, 2014.

9. Henry A. Kissinger, *World Order* (New York: Penguin Press, 2014), 340, 347, 368.

10. See, e.g., Timothy Snyder, *On Tyranny: Twenty Lessons from the Twentieth Century* (New York: Tim Duggan Books, 2017).

11. See, e.g., Jennifer Senior, " 'Richard Nixon,' Portrait of a Thin-Skinned, Media-Hating President," *New York Times*, March 29, 2017; Jennifer Rubin, "End the Nuncs Charade, and Follow the Russian Money," *Washington Post*, March 29, 2017.

12. Jeremiah E. Dittmar, "The Welfare Impact of a New Good: The Printed Book," February 2012 working paper.

13. World Bank Group, *Digital Dividends* (Washington, DC: International Bank for Reconstruction and Development/The World Bank, 2016), 95.

14. World Bank, *Digital Dividends*, 207.

15. World Bank, *Digital Dividends*, xiii, 6.

16. Walter Scheidel, *The Great Leveler: Violence and the History of Inequality from the Stone Age to the Twenty-First Century* (Princeton, NJ: Princeton University Press, 2017).

17. World Bank, *Digital Dividends*, 217.

18. Scale-free networks follow a power law, as the relative likelihood of nodes with very high degree and very low degree is much higher than if edges between nodes were formed at random.

19. Peter Thiel, with Blake Masters, *Zero to One: Notes on Startups, or How to Build the Future* (New York: Crown Business, 2014).

20. In the developing world, cellular service costs vary from nearly $50 a month in Brazil to single digits in Sri Lanka. The price of internet service for a megabit per second is around three hundred times higher in landlocked Chad than in Kenya: World Bank, *Digital Dividends*, 8, 71, 218.

21. World Bank, *Digital Dividends*, 13.

22. Charles Kadushin, "Social Networks and Inequality: How Facebook Contributes to Economic (and Other) Inequality," *Psychology Today*, March 7, 2012, accessed November 4, 2017, https://www.psychologytoday.com/blog/understanding-social-networks/201203/social-networks-and-inequality.

23. Julien Gagnon and Sanjeev Goyal, "Networks, Markets, and Inequality," *American Economic Review* 107, no. 1 (2017): 1–30.

24. Sam Altman, "I'm A Silicon Valley Liberal, and I Traveled across the Country to Interview 100 Trump Supporters—Here's What I Learned," *Business Insider*, February 23, 2017, accessed November 4, 2017, www.businessinsider.com/sam-altman-interview-trump-supporters-2017–2.

25. "As American as Apple Inc.: Corporate Ownership and the Fight for Tax Reform," Penn Wharton Public Policy Initiative, Issue Brief 4, no. 1, accessed November 4, 2017, https://publicpolicy.wharton.upenn.edu/issue-brief/v4n1.php.

26. Sandra Navidi, "How Trumpocracy Corrupts Democracy," *Project Syndicate*, February 21, 2017.

27. Cecilia Kang, "Google, in Post-Obama Era, Aggressively Woos Republicans," *New York Times*, January 27, 2017; Jack Nicas and Tim Higgins, "Silicon Valley Faces Balancing Act between White House Criticism and Engagement," *Wall Street Journal*, January 31, 2017.

28. Issie Lapowsky, "The Women's March Defines Protest in the Facebook Age," *Wired*, January 21, 2017; Nick Bilton, "Will Mark Zuckerberg Be Our Next President?" *Vanity Fair*, January 13, 2017.

29. World Bank, *Digital Dividends*, 221–27.

30. For example, in September 2009, the following pro-Obamacare meme was copied by hundreds of thousands of Facebook users, some of whom (around one in ten) introduced slight mutations to the wording: "No one should die because they cannot afford health care and no one should go broke because they get sick. If you agree please post this as your status for the rest of the day." See Lada A. Adamic, Thomas M. Lenton, Eytan Adar, and Pauline C. Ng, "Information Evolution in Social Networks," February 22–25, 2016, accessed November 4, 2017, https://research.fb.com/wp-content/uploads/2016/11/information_evolution_in_social_networks.pdf.

31. James Stavridis, "The Ghosts of Religious Wars Past Are Rattling in Iraq," *Foreign Policy*, June 17, 2014.

32. Charles S. Maier, *Leviathan 2.0: Inventing Modern Statehood* (Cambridge, MA: Belknap Press, 2014).

33. Fareed Zakaria, "America Must Defend Itself against the Real National Security Menace," *Washington Post*, March 9, 2017.

34. Joseph Nye, "Deterrence and Dissuasion in Cyberspace," *International Security* 41, no. 3 (Winter 2016/17): 47.

35. Joshua Cooper Ramo, *The Seventh Sense: Power, Fortune, and Survival in the Age of Networks* (New York: Little, Brown, 2016), 217f.

36. Guido Caldarelli and Michele Catanzaro, *Networks: A Very Short Introduction* (Oxford: Oxford University Press, 2012), 95–98, 104f.

37. Drew Fitzgerald and Robert McMillan, "Cyberattack Knocks Out Access to Websites," *Wall Street Journal*, October 21, 2016; William Turton, "Everything We Know about the Cyberattack That Crippled America's Internet," *Gizmodo*, October 24, 2016.

38. Fred Kaplan, "'WarGames' and Cybersecurity's Debt to a Hollywood Hack," *New York Times*, February 19, 2016.

39. Nye, "Deterrence and Dissuasion."

40. Ken Dilanian, William M. Arkin, Cynthia McFadden, and Robert Windrem, "U.S. Govt. Hackers Ready to Hit Back If Russia Tries to Disrupt Election," NBC, November 4, 2016.

41. Nathan Hodge, James Marson, and Paul Sonne, "Behind Russia's Cyber Strategy," *Wall Street Journal*, December 30, 2017.

42. For the most recent WikiLeaks release, see Zeynep Tufekci, "The Truth about the WikiLeaks C.I.A. Cache," *New York Times*, March 9, 2017.

43. Bonnie Berkowitz, Denise Lu, and Julie Vitkovskaya, "Here's What We Know So Far about Team Trump's Ties to Russian Interests," *Washington Post*, March 31, 2017.

44. Nye, "Deterrence and Dissuasion," 44–52, 63–67.

45. Mark Galeotti, "Crimintern: How the Kremlin Uses Russia's Criminal Networks in Europe," European Council on Foreign Relations policy brief, April 2017.

46. Anne-Marie Slaughter, "How to Succeed in the Networked World," *Foreign Affairs*, November/December 2016: 80.

47. Slaughter, "How to Succeed," 84f.; Slaughter, *The Chessboard and the Web*.

48. Slaughter, "How to Succeed," 86.

49. Slaughter, *The Chessboard and the Web*.

50. Ramo, *Seventh Sense*, 182.

51. Ramo, *Seventh Sense*, 233.

52. Ramo, *Seventh Sense*, 153.

53. Nassim Nicholas Taleb, *Antifragile: Things That Gain from Disorder* (New York: Random House, 2012).

54. Samuel Arbesman, *Overcomplicated: Technology at the Limits of Comprehension* (New York: Current, 2016).

55. Caldarelli and Catanzaro, *Networks*, 97.

56. Daniel Martin, "Shaming of Web Giants," *Daily Mail*, March 15, 2017.

57. Guy Chazan, "Germany Cracks Down on Social Media over Fake News," *Financial Times*, March 14, 2017.

58. "European Unicorns 2016: Survival of the Fittest," GP Bullhound.

59. Adam Satariano and Aoife White, "Silicon Valley's Miserable Euro Trip Is Just Getting Started," *Bloomberg Business Week*, October 20, 2016; Mark Scott, "The Stakes Are Rising in Google's Antitrust Fight with Europe," *New York Times*, October 30, 2016; Philip Stephens, "Europe Rewrites the Rules for Silicon Valley," *Financial Times*, November 3, 2016.

60. Jack Goldsmith and Tim Wu, *Who Controls the Internet? Illusions of a Borderless World* (Oxford and New York: Oxford University Press, 2008), 5ff.

61. Bethany Allen-Ebrahimian, "The Man Who Nailed Jello to the Wall," *Foreign Policy*, June 29, 2016.

62. Debora L. Spar, *Ruling the Waves: Cycles of Discovery, Chaos, and Wealth from the Compass to the Internet* (Orlando, FL: Harcourt, 2003), 381.

63. Guobin Yang, "China's Divided Netizens," *Berggruen Insights* 6 (October 21, 2016).

64. Goldsmith and Wu, *Who Controls the Internet?*, 96.

65. Emily Parker, "Mark Zuckerberg's Long March to China," *MIT Technology Review*, October 18, 2016; Alyssa Abkowitz, Deepa Seetharaman, and Eva Dou, "Facebook Is Trying Everything to Re-enter China—and It's Not Working," *Wall Street Journal*, January 30, 2017.

66. Mary Meeker, "Internet Trends 2016—Code Conference," Kleiner Perkins Caufield Byers, June 1, 2016, 170f.

67. William C. Kirby, Joycelyn W. Eby, Shuang L. Frost, and Adam K. Frost, "Uber in China: Driving in the Gray Zone," Harvard Business School, Case 9-316-135, May 2, 2016, 12.

68. William Kirby, "The Real Reason Uber Is Giving Up in China," *Harvard Business Review*, August 2, 2016.

69. See, e.g., Eric X. Li, "Party of the Century: How China Is Reorganizing for the Future," *Foreign Affairs*, January 10, 2014.

70. Franziska Barbara Keller, "Networks of Power: Using Social Network Analysis to Understand Who Will Rule and Who Is Really in Charge in the Chinese Communist Party" (PhD diss., New York University, 2015), 32; Franziska Barbara Keller, "Moving beyond Factions: Using Social Network Analysis to Uncover Patronage Networks among Chinese Elites," *Journal of East Asian Studies* 16, no. 1 (March 2016): 22.

71. Cheng Li, *Chinese Politics in the Xi Jinping Era: Reassessing Collective Leadership* (Washington, DC: Brookings Institution, 2016), 332, 347f.

72. Jessica Batke and Matthias Stepan, "Party, State and Individual Leaders: The Who's Who of China's Leading Small Groups," Mercator Institute for China Studies, 2017; Li-Wen Lin and Curtis J. Milhaupt, "Bonded to the State: A Network Perspective on China's Corporate Debt Market," *Journal of Financial Regulation* 3, no. 1 (March 2017).

73. "Chinese Censors' Looser Social Media Grip 'May Help Flag Threats,'" *South China Morning Post*, February 13, 2017.

74. "Visualizing China's Anti-corruption Campaign," *ChinaFile*, January 21, 2016.

75. "Big Data, Meet Big Brother: China Invents the Digital Totalitarian State," *The Economist*, December 17, 2016.

76. Nick Szabo, "Money, Blockchains, and Social Scalability," *Unenumerated*, February 9, 2017.

77. Szabo, "Money, Blockchains, and Social Scalability."

78. Andrew G. Haldane, "A Little More Conversation, a Little Less Action," speech given at the Federal Reserve Bank of San Francisco Macroeconomics and Monetary Policy Conference, March 31, 2017.

79. David McGlauflin, "How China's Plan to Launch Its Own Currency Might Affect Bitcoin," *Crypto Coins*, January 25, 2016; "China Is Developing Its Own Digital Currency," Bloomberg News, February 23, 2017. Details of the PBOC plan are at www.cnfinance.cn/magzi/2016-09/01-24313.html and www.cnfinance.cn/magzi/2016–09/01–24314.html, both accessed November 4, 2017.

80. For a suggestive comparison with the Renaissance, see Ian Goldin and Chris Kutarna, *Age of Discovery: Navigating the Risks and Rewards of Our New Renaissance* (New York: St. Martin's Press, 2016).

81. Francis Heylighen and Johan Bollen, "The World-Wide Web as a Super-Brain: From Metaphor to Model," in *Cybernetics and Systems*, edited by Robert Trappl (Vienna: Austrian Society for Cybernetics, 1996).

82. Michael Dertouzos, *What Will Be: How the New World of Information Will Change Our Lives* (New York: HarperEdge, 1997).

83. Robert Wright, *Nonzero: The Logic of Human Destiny* (New York: Vintage, 2001), 198.

84. N. Katherine Hayles, "Unfinished Work: From Cyborg to Cognisphere," *Theory, Culture & Society* 23, no. 159 (2006): 164.

85. Ian Tomlin, *Cloud Coffee House: The Birth of Cloud Social Networking and Death of the Old World Corporation* (Cirencester, UK: Management Books, 2000, 2009), 55.

86. Tomlin, *Cloud Coffee House*, 223.

87. Fred Spier, *Big History and the Future of Humanity* (Malden, MA, and Oxford: Wiley-Blackwell, 2011), 138–83.

88. John Naughton, *From Gutenberg to Zuckerberg: What You Really Need to Know about the Internet* (London: Quercus, 2012), 207, 236.

89. Robert J. Gordon, *The Rise and Fall of American Growth: The U.S. Standard of Living since the Civil War* (Princeton, NJ: Princeton University Press, 2016).

90. Daron Acemoglu and Pascual Restrepo, "Robots and Jobs: Evidence from US Labor Markets," NBER working paper no. 23285 (March 2017); World Bank, *Digital Dividends*, 23, 131.

91. Bryan Caplan, "The Totalitarian Threat," in *Global Catastrophic Risks*, ed. Nicholas Bostrom and Milan M. Cirkovic (Oxford: Oxford University Press, 2008), 504–18.

92. For a historically based prediction of an upsurge in violence in the United States, see Peter Turchin, *Ages of Discord: A Structural-Demographic Analysis of American History* (Chaplin, CT: Beresta Books, 2016).

93. Nicholas Bostrom, *Superintelligence: Paths, Dangers, Strategies* (Oxford, UK: Oxford University Press, 2014).

94. James C. Scott, *Two Cheers for Anarchism: Six Easy Pieces on Autonomy, Dignity, and Meaningful Work and Play* (Princeton, NJ: Princeton University Press, 2012).

95. Niall Ferguson, "Donald Trump's New World Order," *The American Interest* 12, no. 4 (November 21, 2016): 37–47.

7. Governance from an International Perspective

1. Farhad Manjoo, "Social Media's Globe-Shaking Power," *New York Times,* November 16, 2016.

2. "Will the Internet of Things Be a Cybersecurity Disaster?" Council on Foreign Relations roundtable discussion, May 25, 2017.

3. Dinner discussion with Governor John Kasich in Munich, Germany, February 18, 2017.

4. Harriet Agnew and Jim Brunsden, "France Urges 'New Momentum' in Taxation of US Tech Groups," *Financial Times*, August 10, 2017.

5. Margrethe Vestager, "A Healthy Democracy in a Social Media Age," speech to ALL for Democracy, Brussels, June 7, 2017.

6. Margrethe Vestager, "Fighting for European Values in a Time of Change," speech, Leiden University, Netherlands, June 14, 2017.

7. Natalia Drozdiak and Sam Schechner, "Tech Firms Gird for New EU Privacy Law," *Wall Street Journal*, December 13, 2015.

8. Guy Chazan, "Rise of Refugee 'Fake News' Rattles German Politics," *Financial Times*, February 14, 2017.

9. David Bond and Duncan Robinson, "European Commission Fires Warning at Facebook over Fake News," *Financial Times*, January 30, 2017.

10. Nathan Hodge, James Marson, and Paul Sonne, "Behind Russia's Cyber Strategy," *Wall Street Journal*, December 30, 2016.

11. Stefan Wagstyl, "Berlin Braces for Russian Meddling before September Election," *Financial Times*, July 4, 2017.

12. Sam Jones, "Russia Mobilizes an Elite Band of Cyber Warriors," *Financial Times*, February 23, 2017.

13. Stefan Wagstyl, "Germans Detect Hand of Russia as Political Cyber War Escalates," *Financial Times*, December 29, 2016.

14. Andrew Higgins, "Intent on Unsettling E.U., Russia Taps Foot Soldiers from the Fringe," *New York Times*, December 24, 2016.

15. Jason Horowitz, "Spread of Fake News Provokes Anxiety in Italy," *New York Times*, December 2, 2016.

16. Aurelien Breeden, Sewell Chan, and Nicole Perlroth, "Macron Campaign Says It Was Target of 'Massive' Hacking Attack," *New York Times*, May 5, 2017.

17. Thomas Grove, "Estonia Leads the Way in NATO's Cyberdefense," *Wall Street Journal*, April 30, 2017.

18. Martin Libicki, "China Developing Cyber Capabilities to Disrupt U.S. Military Operations," *The Cipher Brief*, April 2, 2017.

8. Governance and the American Presidency

1. Theodore White, *The Making of the President 1960* (New York: Atheneum, 1961), 222.

2. *Federalist* no. 70 (Alexander Hamilton), "The Executive Department Further Considered," March 18, 1788, accessed August 13, 2016, http://avalon.law.yale.edu/18th_century/fed70.asp.

3. Woodrow Wilson, *Congressional Government: A Study in American Politics* (New York: Houghton Mifflin, 1885), 318.

4. Woodrow Wilson, speech at Columbia University, 1908, Teaching AmericanHistory.org, accessed August 13, 2016, http://teachingamerican history.org/library/document/constitutional-government-chapter-iii-the -president-of-the-united-states.

5. Jeffrey K. Tulis, *The Rhetorical Presidency* (Princeton, NJ: Princeton University Press, 1987), 138–44.

6. Edmund Burke, "Speech on Conciliation with the Colonies," March 22, 1775, accessed August 11, 2017, http://press-pubs.uchicago.edu/founders/documents/v1ch1s2.html.

7. Alexis de Tocqueville, *Democracy in America*, vol. 2, part 2, chap. 28 (1835).

8. Terry Moe and William Howell, *Relic: How Our Constitution Undermines Effective Government—And Why We Need a More Powerful Presidency* (New York: Basic Books, 2016).

9. Technological Change and Language

1. Thomas Kuhn, *The Structure of Scientific Revolutions* (Chicago: University of Chicago Press, 2012).

2. Carlos Eire, *Reformations: The Early Modern World, 1450–1650* (New Haven, CT: Yale University Press, 2017). Reviewed in the *Times Literary Supplement* by Charlotte Methuen, May 5, 2017.

3. "As our case is new, so we must think anew and act anew." Abraham Lincoln, Second Annual Message, December 1, 1862.

4. Kant's term.

5. David P. Jordan, *Gibbon and His Roman Empire* (Champaign: University of Illinois Press, 1971), 79, 121.

6. Gibbon, *Memoirs of My Life and Writings* (London: Penguin, 1984, first published 1796).

7. J. H. Parry, *The Age of Reconnaissance: Discovery, Exploration, and Settlement, 1450–1650* (New York: World Publishing Company, 1963).

8. Adam Smith, *The Wealth of Nations*, book 6, chapter 7, part 3 (New York: Modern Library, 1994), 675.

9. Emerson's ladder recalls the "Ladder of Love" in Plato's *Symposium* and other, later conceptions in the liberal arts that attempt to connect the low real and the high ideal in order to satisfy the desire for "meaning" in some sort of metaphysical outcome.

10. A term from Emerson to describe Samuel Coleridge.

11. Note also Dante's reference at the end of *Paradiso* to the voyage of Jason and the Argonauts, the first-ever ship to venture beyond coasting to sail the high seas. Neptune, seeing the shadow of the *Argo*, is infuriated by this

transgression. What is the relation of his words, the shadow, and the ship? Also see William Carlos Williams's famous line, "No ideas but in things."

12. Mary Shelley, *Frankenstein*, ed. J. Paul Hunter, Norton Critical Editions, 2nd ed. (New York: W. W. Norton, 2012), ix.

13. Hunter, *Frankenstein*, 5.

14. Lawrence Lipking, "Frankenstein, The True Story; or Rousseau Judges Jean-Jacques," in Hunter, *Frankenstein*, 416–33.

15. Peter Brooks, "What Is a Monster?" in Hunter, *Frankenstein*, 368–90.

16. Charles Hill, "Man, the Religious Animal," *Defining Ideas*, May 3, 2017.

17. Ranajit Guha, *History at the Limit of World-History* (New York: Columbia University Press, 2002).

18. George Levine, introduction to *The Origin of Species* by Charles Darwin (New York: Barnes and Noble Classics, 2004); Philip Appleman, ed., *Darwin*, Norton Critical Editions, 2nd ed. (New York: W. W. Norton, 1970) (jacket copy).

19. Thomas Nagel, *Mind & Cosmos: Why the Materialist Neo-Darwinian Conception of Nature Is Almost Certainly False* (Oxford: Oxford University Press, 2012), 10–11. See also Marilynne Robinson, "Darwinism," in *The Death of Adam: Essays on Modern Thought* (New York: Picador, 1998).

20. See Stanley Fish, "How Hobbes Works," in *Versions of Antihumanism: Milton and Others* (Cambridge, UK: Cambridge University Press, 2002).

21. Albert W. Alschuler, *Law without Values: The Life, Work, and Legacy of Justice Holmes* (Chicago: University of Chicago, 2000).

22. Nathaniel Berman, "Modernism, Nationalism, and the Rhetoric of Reconstruction," *Yale Journal of Law and the Humanities* 4, no. 2 (1992): 351.

23. Bruce A. Kimball, *Orators and Philosophers: A History of the Idea of Liberal Education* (New York: The College Board, 1995).

24. *Understanding Islam and the Muslims*, Islamic Affairs Department, Embassy of Saudi Arabia, Washington, DC, 1989.

25. Kenneth Cragg, *The House of Islam* (Dickenson, 1969), 32; Jean-Pierre Peroncel-Hugoz, *The Raft of Mohammed: Social and Human Consequences of the Return to Traditional Religion in the Arab World* (St. Paul, MN: Paragon House, 1988), 157–59.

26. Mohammed Marmaduke Pickthall, *The Meaning of the Glorious Koran* (Denver, CO: Mentor Books, 1955), xxix.

27. A. A. Duri, *The Rise of Historical Writing among the Arabs* (Princeton, NJ: Princeton University Press, 1984), 18.

28. In accordance with the potential noted by Bernard Lewis, "Propaganda in the Pre-modern Middle East," in *From Babel to Dragomans: Interpreting the Middle East* (Oxford: Oxford University Press, 2004), 79, 84, 85.

29. Lewis, *From Babel to Dragomans,* 86.

30. Dan Diner, *Lost in the Sacred: Why the Muslim World Stood Still* (Princeton, NJ: Princeton University Press, 2009), 69–73.

31. Charles Taylor, *The Language Animal: The Full Shape of the Human Linguistic Capacity* (Cambridge, MA: Harvard University Press, 2016).

32. Louis Menand, introduction to the 75th anniversary edition of *Civilization and Its Discontents,* by Sigmund Freud (New York: W. W. Norton, 2005), 14.

33. Samuel P. Huntington, *Political Order in Changing Societies* (New Haven, CT: Yale University Press, 1968).

34. And, says Tocqueville, it is women who maintain religion. Indeed, he says, America's success overall is owing to "the superiority of their women."

35. Paul M. Churchland, *Matter and Consciousness* (Cambridge, MA: MIT Press, 1984).

ACKNOWLEDGMENTS

G reat help in assembling this book was provided by David Fedor and Susan Southworth. Hoover Institution's Denise Elson, Chris Dauer, and Shana Farley, along with Janet Smith, Linda Hernandez, and Kelly Russo of Hoover's events staff, were superb facilitators of the conference during which our authors presented their work. A final panel discussion among the authors was open to the Stanford University community and served as the public unveiling of Hauck Auditorium in Hoover Institution's new Traitel Building. Contributions to the success of these events and to the production of this book are gratefully acknowledged.

CONTRIBUTORS

William Drozdiak, a career transatlanticist, is a nonresident senior fellow in foreign policy at the Brookings Institution's Center for the United States and Europe and senior adviser for Europe and Eurasia at McLarty Associates. He led the American Council on Germany, an organization devoted to cooperation between the United States and Europe, for ten years and previously served in Brussels as the founding executive director of the Transatlantic Center of the German Marshall Fund of the United States. He worked for two decades as an editor and foreign correspondent for the *Washington Post*. He is the author of *Fractured Continent: Europe's Crises and the Fate of the West* (Norton, 2017).

Admiral James O. Ellis, Jr., is the Annenberg Distinguished Fellow at the Hoover Institution. In 2012 he retired as president and chief executive officer of the Institute of Nuclear Power Operations (INPO). He earlier completed a distinguished thirty-nine-year Navy career as commander of the US Strategic Command. In this role, he was responsible for the global command and control of US strategic and space forces, reporting directly to the secretary of defense. He earlier commanded the nuclear-powered aircraft carrier USS *Abraham Lincoln* and in 1996 served as a carrier battle group commander in the Taiwan Straits crisis, following selection to rear admiral. Senior shore assignments included commander in chief, US Naval Forces, Europe, and commander in chief, Allied Forces, Southern Europe. A 1969 graduate of the US Naval Academy, Ellis holds a master's degree in aerospace engineering from Georgia Tech, served multiple tours as a Navy fighter pilot, and is a graduate of the Navy Test Pilot School and the Navy Fighter Weapons School (Top Gun). Ellis was recently elected to the National Academy of Engineering.

Niall Ferguson is the Milbank Family Senior Fellow at the Hoover Institution and a senior fellow of the Center for European Studies, Harvard. He is also a visiting professor at Tsinghua University, Beijing, and the Diller–von Furstenberg Family Foundation Distinguished Scholar at the Nitze School of Advanced International Studies in Washington, DC. He has written fourteen books, including *The House of Rothschild*, *Empire*, *The War of the World*, *The Ascent of Money*, *The Great Degeneration*, and *Kissinger, 1923–1968: The Idealist*. His 2011 feature-length film *Kissinger* won the New York International Film Festival's prize for best documentary. His PBS series *The Ascent of Money* won the International Emmy for best documentary. His many prizes and awards include the Benjamin Franklin Prize for Public Service (2010), the Hayek Prize for Lifetime Achievement (2012), and the Ludwig Erhard Prize for Economic Journalism (2013). He writes a weekly column for the London *Sunday Times* and the *Boston Globe*.

T. X. Hammes is a senior research fellow at the Institute for National Strategic Studies, National Defense University. His areas of expertise include future conflict, military strategy, and insurgency. He has published fifteen book chapters and more than 120 articles. Hammes has lectured extensively at leading academic and military institutions in the United States and abroad. Prior to his retirement from active duty, Hammes served for thirty years in the US Marine Corps. He graduated with a BS from the US Naval Academy and holds a master's degree in historical research and a doctorate in modern history from Oxford University. He is also a distinguished graduate of the Canadian National Defence College.

Charles Hill is a research fellow and a co-chairman of the Working Group on Islamism and the International Order at the Hoover Institution. Hill is a career minister in the US Foreign Service and served as the executive aide to former US secretary of state George P. Shultz (1985–89) and as special consultant on policy to the secretary-general of the United Nations (1992–96). He is also the Brady-Johnson Distinguished Fellow in Grand Strategy and a senior lecturer in humanities at Yale. Hill has collaborated with former UN secretary-general Boutros Boutros-Ghali on *Egypt's Road to Jerusalem*, a memoir of the Middle East peace negotiations, and *Unvanquished*, about US relations with the United Nations in the post–Cold War period. His book *Grand Strategies: Literature, Statecraft, and World Order* was published by Yale University Press in 2010. Hill's most recent book is *Trial of a Thousand Years: World Order and Islamism* (Hoover Institution Press, 2011).

Jim Hoagland is an Annenberg Distinguished Visiting Fellow at the Hoover Institution and has served in a variety of reporting, editing, and opinion-forming roles at the *Washington Post* since joining the newspaper in 1966. Winner of two Pulitzer Prizes, he became a contributing editor to the *Post* in 2010, after serving two decades as associate editor and chief foreign correspondent. In 2002, the editors of seven leading European newspapers headed a jury that awarded Hoagland the Cernobbio-Europa Prize. He is the author of *South Africa: Civilizations in Conflict*, which was published in 1972. He graduated with an AB in journalism from the University of South Carolina and was elected to Phi Beta Kappa. He did graduate work in Aix-en-Provence, France, and was a Ford Foundation Fellow at Columbia University's School of Journalism.

Raymond Jeanloz is an Annenberg Distinguished Visiting Fellow at the Hoover Institution and a professor of Earth and planetary science and astronomy and a senior fellow at the Miller Institute for Basic Research in Science at the University of California, Berkeley. In addition to his scientific research on the evolution of planetary interiors and properties of materials at high pressures, he works at the interface between science and policy in areas related to national and international security, resources and the environment, and education. He is a former member of the Secretary of State's International Security Advisory Board and currently chairs the National Academy of Sciences Committee on International Security and Arms Control.

David M. Kennedy is the Donald J. McLachlan Professor of History, Emeritus, at Stanford. He received the Dean's Award for Distinguished Teaching in 1988. Reflecting his interdisciplinary training in American studies, which combines the fields of history, literature, and economics, Professor Kennedy's scholarship is notable for its integration of economic and cultural analysis with social and political history. He was awarded the Pulitzer Prize in history in 2000 for *Freedom from Fear: The American People in Depression and War*. He received an AB in history from Stanford University and an MA and PhD from Yale University.

Harley McAdams, formerly of Bell Labs and the Lockheed Missile and Space Co., is a professor emeritus of the Department of Developmental Biology at Stanford University's School of Medicine. His laboratory focuses on understanding the nature of bacterial genetic regulation through the use of a variety of tools: gene expression microarrays, advanced techniques of fluorescence and electron microscopy, and a broad range of computational tools for bioinformatics and modeling. McAdams is a fellow of the American Academy of Microbiologists and in 2009 received the John Scott Award from the City of Philadelphia.

Nicole Perlroth covers cybersecurity and privacy for the *New York Times*. Perlroth was awarded the "Best in Business" prize by the Society of American Business Editors for her coverage of Chinese cyberespionage and was selected as one of the Top Cybersecurity Journalists by the

SANS Institute. She broke the story of the National Security Agency's efforts to crack encryption and helped identify two Chinese military units responsible for thousands of attacks on US institutions. Her 2014 *Times* profile of security blogger Brian Krebs has been optioned by Sony Pictures. She is currently at work on *This Is How They Tell Me the World Ends* (Penguin/Portfolio, forthcoming).

Lucy Shapiro is a professor in the Department of Developmental Biology at Stanford University's School of Medicine, where she holds the Virginia and D. K. Ludwig Chair in Cancer Research; she is also director of the Beckman Center for Molecular and Genetic Medicine. She is a member of the board of advisers of the Pasteur Institute, the Ludwig Institute for Cancer Research, and the Lawrence Berkeley National Laboratory. She founded the anti-infectives discovery company Anacor Pharmaceuticals and is a member of its board of directors. She has received multiple honors, including election to the American Academy of Arts and Sciences and the National Academy of Sciences. She was awarded the 2005 Selman A. Waksman Award from the National Academy of Sciences, the Canadian International 2009 Gairdner Award, the 2009 John Scott Award, and the 2010 Abbott Lifetime Achievement Award.

George Pratt Shultz is the Thomas W. and Susan B. Ford Distinguished Fellow at the Hoover Institution. He has had a distinguished career in government, in academia, and in the world of business. He is one of two individuals who have held four different federal cabinet posts; he has taught at three of this country's great universities; and for eight years he was president of a major engineering and construction company. Shultz was sworn in July 16, 1982, as the sixtieth US secretary of state and served until January 20, 1989. He attended Princeton University, graduating with a BA in economics, whereupon he enlisted in the US Marine Corps, serving through 1945. He later earned a PhD in industrial economics from the Massachusetts Institute of Technology. In 1989, Shultz was awarded the Medal of Freedom, the nation's highest civilian honor.

His most recent book is *Learning from Experience* (Hoover Institution Press, 2016).

Christopher Stubbs is an Annenberg Distinguished Visiting Fellow at the Hoover Institution. He is also a professor of physics and of astronomy at Harvard University and previous chair of Harvard's Physics Department. His research interests lie at the intersection of cosmology, particle physics, and gravitation, with more than four hundred publications to date. He also has a strong interest in national security. Stubbs is a member of JASON, a group of scientists and engineers who provide technical advice to government agencies on national security issues. He also serves on the technical advisory group for the US Senate Select Committee on Intelligence. Stubbs is a fellow of the American Physical Society, a recipient of the National Academy of Sciences Award for Initiative in Research and the NASA Achievement Medal, and a co-recipient of the Gruber Foundation Cosmology Prize.

John B. Taylor is the George P. Shultz Senior Fellow in Economics at the Hoover Institution and the Mary and Robert Raymond Professor of Economics at Stanford University. He chairs the Hoover Working Group on Economic Policy and is director of Stanford's Introductory Economics Center. Taylor's fields of expertise are monetary policy, fiscal policy, and international economics, subjects about which he has widely authored both policy and academic texts. Taylor served as senior economist on President Ford's and President Carter's Council of Economic Advisers, as a member of President George H. W. Bush's Council of Economic Advisers, and as a senior economic adviser to numerous presidential campaigns. He was a member of the Congressional Budget Office's Panel of Economic Advisers, and from 2001 to 2005 he served as undersecretary of the treasury for international affairs. Taylor received a BA in economics summa cum laude from Princeton University in 1968 and a PhD in economics from Stanford University in 1973.

James Timbie is an Annenberg Distinguished Visiting Fellow at the Hoover Institution. As senior adviser at the State Department from 1983 to 2016, Timbie played a central role in the negotiation of the INF and START nuclear arms reductions agreements, the purchase from Russia of enriched uranium extracted from dismantled nuclear weapons for use as fuel to produce electricity in the United States, and the establishment of an international enriched uranium fuel bank. Most recently he was the lead US expert in the negotiation of the nuclear agreement with Iran. He retired from the State Department in 2016. He has a PhD in physics from Stanford University and from 1971 to 1983 was a scientist at the Arms Control and Disarmament Agency.

September 2017 conference participants provided additional reflections:

Persis Drell is the Stanford University provost and former dean of the Stanford School of Engineering.

Karl Eikenberry is a fellow at the Stanford University Freeman Spogli Institute, former U.S. ambassador to Afghanistan, and retired US Army lieutenant general.

Ernest Moniz is CEO of the Nuclear Threat Initiative, MIT professor of physics post-tenure, and former US secretary of energy.

Sam Nunn is co-chair of the Nuclear Threat Initiative and former US senator from Georgia.

William J. Perry is a Stanford University professor emeritus and former US secretary of defense.

Burton Richter is a Stanford University professor emeritus, director emeritus at the SLAC National Accelerator Laboratory, and 1976 winner of the Nobel Prize in physics.

Kori Schake is a distinguished research fellow at the Hoover Institution and former staff member of the White House National Security Council.

Thomas F. Stephenson is co-chair of the Hoover Institution's Shultz-Stephenson Task Force on Energy Policy and a partner at Sequoia Capital.

William Swing is the founder of the United Religions Initiative and retired Episcopal bishop of California.

Tsunehiko Yanagihara is executive vice president of Mitsubishi Corporation Americas based in Silicon Valley.

INDEX